T0192509

HIDDEN MARKOV MODELS
Theory and Implementation using MATLAB®

João Paulo Coelho
Instituto Politécnico de Bragança and
CeDRI - Research Centre in Digitalization and Intelligent Robotics
Portugal

Tatiana M. Pinho
INESC TEC Technology and Science
Portugal

José Boaventura-Cunha
Universidade de Trás-os-Montes e Alto Douro and
INESC TEC Technology and Science
Portugal

CRC Press
Taylor & Francis Group
Boca Raton London New York

CRC Press is an imprint of the
Taylor & Francis Group, an **informa** business

A SCIENCE PUBLISHERS BOOK

Cover credit: Lourenço Rafael Cardoso Ribeiro

CRC Press
Taylor & Francis Group
6000 Broken Sound Parkway NW, Suite 300
Boca Raton, FL 33487-2742

First issued in paperback 2021

© 2019 by Taylor & Francis Group, LLC
CRC Press is an imprint of Taylor & Francis Group, an Informa business

No claim to original U.S. Government works

Version Date: 20190401

ISBN 13: 978-0-367-77934-4 (pbk)
ISBN 13: 978-0-367-20349-8 (hbk)

Library of Congress Cataloging-in-Publication Data

Names: Coelho, João Paulo (Engineer), author. | Pinho, Tatiana M.,
    author. | Boaventura-Cunha, José, author.
Title: Hidden Markov models : theory and implementation using Matlab / João
    Paulo Coelho (Instituto Politécnico de Bragança-Escola Superior de
    Tecnologia e Gestão, Bragança, Portugal), Tatiana M. Pinho (INESC TEC
    Technology and Science, Porto, Portugal), José Boaventura-Cunha
    (Universidade de Trás-os-Montes e Alto Douro, Vila Real, Portugal).
Description: Boca Raton, FL : CRC Press, 2019. | "A science publishers book."
    | Includes bibliographical references and index.
Identifiers: LCCN 2019010285 | ISBN 9780367203498 (hardback)
Subjects: LCSH: Markov processes. | Hidden Markov models. | Stochastic
    processes | Markov processes--Data processing. | Hidden Markov
    models--Data processing. | Stochastic processes--Data processing. | MATLAB.
Classification: LCC QA274.7 .C635 2019 | DDC 519.2/33--dc23
LC record available at https://lccn.loc.gov/2019010285

Visit the Taylor & Francis Web site at
http://www.taylorandfrancis.com

and the CRC Press Web site at
http://www.crcpress.com

Preface

"We are all just cogs in a machine, doing what we were always meant to do, with no actual volition."

—Baron d'Holbach

The quotation above is attributed to Paul-Henri Thiry, Baron d'Holbach, a XVIII century French philosopher. The statement describes, in a very dogmatic and absolute way, the position of d'Holbach regarding the subject of free will versus determinism. This dichotomy of thought is not recent and discussions on this topic date back as far as ancient Greek philosophers. In one hand, it is our internal belief that we are able to choose between distinct courses of action without being driven by any higher force. We don't feel that the future is pre-determined from the beginning of time. Instead, its structure is flexible by nature and can be bent by each and everyone one of us. On the other hand, it is our understanding that the universe is deterministic and, hence, only one course of events is possible. That is, in the limit, we don't have any control on our actions due to the fact that the future is correlated with the past since the beginning of time. Note that this reasoning is in line with the Newtonian interpretation of nature. According to it, the laws that govern everything around us, are deterministic[1]. If two exactly identical rocks are thrown, at different time instants, in the same direction, with the same initial velocity, angle and so on then, according to the Newtonian laws of physics, both will reach the same distance and this distance can be accurately predicted. Now, even if determinism seems counterintuitive when applied to living organisms, it fits better with our knowledge of reality than the concept of free will. Are we so different from rocks?

Suppose that, at a given time instant, we are able to reboot the universe to some arbitrary state. Let this state be, for example, the one derived at the exact

---

[1] Of course, if we are dealing with the universe at the quantum level, things behave in a very different way. For example the existence of quantum superposition is a reality for subatomic entities. However, quantum effects are diluted when dealing with things at higher levels in the matter scale.

time instant when you have been conceived by your parents. Is it then rational to think that, all the outcomes that follow immediately after this reboot will be the same? That is, you will be created and everything following this event will lead to the exactly same world as the one we are living on right now?

Let's return to our ability to make choices. In our everyday lives, we make hundreds of choices and it seems that we are in charge of all the options made. Can we say that, under the exactly same stimuli, we are able to choose otherwise? Well, if reasoning is an outcome of the brain, and if the brain is a large network of neurons and, at the same time, the neurons always fire in a deterministic fashion whenever submitted to the same stimulus, then probably we are unable to choose differently from the initially picked option. For the sake of argument, let's then assume that free will is unattainable. Can we then predict with exactitude what each and every one of us will do under some condition? This is similar to ask if, in practice, we can predict, with infinite accuracy, where the thrown rock will fall.

The problem is that we must know how each entity will behave under some stimuli, know its initial state or be able to observe and compute a state space which can be prohibitively large and unreachable. So, in short, it is very unlikely that we are able to know, with zero error, what will be the outcome of some event. When asked about the landing point of a thrown rock we can answer something like—"I'm 99% sure that the rock will land at coordinates $(x, y) \pm (\Delta x, \Delta y)$". This means that, due to the universe complexity and processes intertwining, the best thing we are able to do, with more or less certainty, is to guess...

This book deals with this nondeterminism and provides a consistent framework on how to handle a class of very important mathematical models known as hidden Markov models. The main goal is to illustrate the mathematics behind these types of models using intuition and a high level computer language. Having said that, let's begin our voyage.

João Paulo Coelho
Tatiana M. Pinho
José Boaventura-Cunha

Contents

Glossary

k	Index normally associated to the Markov model switching time instant.
m	Number of hidden states in a hidden Markov model.
n	Number of observable states in a hidden Markov model.
l	Index usually associated to the current iteration number.
r_j	In a discrete hidden Markov model, there is a state associated to each observation. Hence r_j regards the observable state j for $j = 1, ..., n$.
λ_k	Active observable state at time instant k. At a given time instant k, λ_k regards one of the n observable states. That is, $\lambda_k \in \{r_1, \cdots, r_n\}$.
Λ_N	Sequence of the active observable states from $k = 1$ to $k = N$. That is, $\Lambda_N = \lambda_1, \cdots, \lambda_N$.
Λ_k	Sequence of the active observable states from 1 to k, where $k \leq N$. That is, $\Lambda_k = \lambda_1, \cdots, \lambda_k$.
$\Lambda_{a \to b}$	Sequence of the active observable states from a to b, for $a, b \in \mathbb{N}$ as long as $b > a$. That is, $\Lambda_{a \to b} = \lambda_a, \cdots, \lambda_b$.
q_k	Hidden state active at time instant k.
s_i	The ith hidden state where i can be any number between 1 and m.
c_i	Probability of having s_i as the starting active hidden state. Also referred as the priors.
$\mathbf{c}$	Priors vector with dimension $1 \times m$ with the format: $\mathbf{c} = [c_1 \cdots c_m]$.
a_{ij}	Transition probability from hidden state s_i to s_j. That is, $P(q_k = s_j | q_{k-1} = s_i)$. Also refers to the ith row, jth column element of matrix $\mathbf{A}$.

A	Probability transition matrix with the following structure:

$$\mathbf{A} = [a_{ij}] = \begin{bmatrix} a_{11} & \cdots & a_{1m} \\ \vdots & & \vdots \\ a_{m1} & \cdots & a_{mm} \end{bmatrix}$$

$b_j(\lambda_k)$ — Probability of attaining the observable state k given that the active hidden state is s_j.

b_{ji} — Probability of observing the state r_i given that the active hidden state is s_j. That is, $b_{ji} = P(\lambda_k = r_i | q_k = s_j)$. This is also referred as the emission probability.

B — Emission probabilities matrix described as:

$$\mathbf{B} = [b_{ji}] = \begin{bmatrix} b_{11} & \cdots & b_{1n} \\ \vdots & & \vdots \\ b_{m1} & \cdots & b_{mn} \end{bmatrix}$$

Θ — Parameters set (**A**, **B**, **c**) that fully characterize a particular hidden Markov model. That is, $\Theta \rightarrow (\mathbf{A}, \mathbf{B}, \mathbf{I})$. In the context of Gaussian mixtures or continuous hidden Markov models, it can also refer to $\Theta \rightarrow (\mu, \Sigma)$.

$\alpha_k(i)$ — Forward probability. The probability of observing the sequence $\lambda_1 \cdots \lambda_k$ and the active hidden state at k be s_i. That is,

$$\alpha_k(i) = P(\Lambda_k \wedge q_k = s_i)$$

u_k — Scale factor.

$\bar{\alpha}_k(i)$ — Scaled version of $\alpha_k(i)$. That is, $\bar{\alpha}_k(i) = u_k \cdot \alpha_k(i)$.

$\boldsymbol{\alpha}_k$ — The forward probability vector at time instant k with the structure $\boldsymbol{\alpha}_k = [\alpha_k(1) \cdots \alpha_k(m)]$.

$\mathcal{A}$ — Forward probability matrix obtained from $\alpha_k(i)$ for $k = 1, \cdots, N$ and $i = 1, \cdots, m$ represented as:

$$\mathcal{A} = \begin{bmatrix} \boldsymbol{\alpha}_1 \\ \vdots \\ \boldsymbol{\alpha}_N \end{bmatrix} = \begin{bmatrix} \alpha_1(1) & \cdots & \alpha_1(m) \\ \vdots & & \vdots \\ \alpha_N(1) & \cdots & \alpha_N(m) \end{bmatrix}$$

$\beta_k(i)$ — Backward probability. It regards the probability of observing the partial sequence $\{\lambda_{k+1}, \cdots, \lambda_N\}$ given that the active hidden state at time k is s_i. That is, $\beta_k(i) = P(\Lambda_{k+1 \rightarrow N} | q_k = s_i)$.

v_k	Scale factor.
$\bar{\beta}_k(i)$	Scaled version of $\beta_k(i)$. That is, $\bar{\beta}_k(i) = v_k \cdot \beta_k(i)$.
β_k	The backward probability vector at time instant k with the structure $\beta_k = [\beta_k(1) \cdots \beta_k(m)]$.
$\mathcal{B}$	Backward probability matrix obtained from $\beta_k(i)$ for $k = 1, \cdots, N$ and $i = 1, \cdots, m$ represented as:

$$\mathcal{B} = \begin{bmatrix} \beta_1 \\ \vdots \\ \beta_N \end{bmatrix} = \begin{bmatrix} \beta_1(1) & \cdots & \beta_1(m) \\ \vdots & & \vdots \\ \beta_N(1) & \cdots & \beta_N(m) \end{bmatrix}$$

δ_i	Hidden state index pointed out by q_i.
$Q_l = \{q_1, \cdots, q_k\}$	Hidden states sequence up to time k.
$\mathcal{L}(\Theta\|\Lambda_k)$	Likelihood of a generic m hidden states Markov model conditioned to a set of k observations.
$\rho_j(i)$	Maximum probability of obtaining the observable sequence from Λ_j supposing that, at $k = j$, the active hidden state is s_i. That is $\rho_j(i) = \max\{P(\lambda_1 \cdots \lambda_j \wedge q_j = s_i)\}$.
$\rho'_j(i)$	Normalized version of $\rho_j(i)$.
$\psi_j(i)$	Works as a pointer and defines, during iteration j, which hidden states will lead, at $k = j - 1$, to a larger ρ_k value.
$\gamma_k(i)$	Probability to have s_i as the active hidden state given the observation sequence Λ_N. That is, $\gamma_k(i) = P(q_k = s_i\|\Lambda_N)$.
Γ	Probability matrix regarding $\gamma_k(i)$ for $k = 1, \cdots, N$ and $i = 1, \cdots, m$ with the structure:

$$\Gamma = \begin{bmatrix} \gamma_1(1) & \cdots & \gamma_1(m) \\ \vdots & & \vdots \\ \gamma_N(1) & \cdots & \gamma_N(m) \end{bmatrix}$$

v_i	Expected number of visits to hidden state s_i along N observations.
$\mathbf{V}$	Matrix of v_i for $i = 1, \cdots, m$.

$$\mathbf{V} = \begin{bmatrix} v_1 & \cdots & v_1 \\ \vdots & & \vdots \\ v_m & \cdots & v_m \end{bmatrix}$$

τ_i	Number of times a transition from state s_i is observed.
$\xi_k(i, j)$	Given the sequence Λ_N, probability of having s_i as the active hidden state at time instant k and having s_j active at $k + 1$. That is, $\xi_k(i, j) = P(q_k = s_i \wedge q_{k+1} = s_j | \Lambda_N)$.
τ_{ij}	Number of times the transition from hidden state s_i to s_j occurs along N observations.
ω_{ij}	Number of times that, being at state s_i, the state r_j is observed.
Q^*	Refers to the most probable sequence of hidden states. That is, $Q^* = \{q_0^* \, q_1^* \cdots q_N^*\}$.
$\Omega_a(\xi, \varepsilon)$	Partial derivative of $\mathcal{L}(\Theta | \Lambda_k)$ in order to $a_{\xi\varepsilon}$.
$\bar{\Omega}_a(\xi, \varepsilon)$	Scaled version of $\Omega_a(\xi, \varepsilon)$.
$\Phi_k(\xi, \varepsilon, i)$	Partial derivative of α_k regarding $\partial a_{\xi\varepsilon}$.
$\bar{\Phi}_k(\xi, \varepsilon, i)$	Scaled version of $\Phi_k(\xi, \varepsilon, i)$.
$\Psi_k(\xi, \varepsilon, i)$	Partial derivative of $\alpha_k(i)$ in order to $b_{\xi\varepsilon}$.
$\bar{\Psi}_k(\xi, \varepsilon, i)$	Scaled version of $\Psi_k(\xi, \varepsilon, i)$.
η_l	Learning rate at iteration l.
$L(\cdot)$	Lagrangian function.
$\mathbf{H}$	Hessian matrix.
$\nabla f(\mathbf{x})|_{\mathbf{x} = \mathbf{x}_l}$	Gradient of $f(\mathbf{x})$ around $\mathbf{x} = \mathbf{x}_l$.
M	Number of Gaussian functions that compose a given Gaussian mixture function.
w_{ij}	Gaussian mixture weight coefficients.
$\mathbf{w}_i$	Gaussian mixture weights, with structure $\mathbf{w}_i = [w_{i1} \cdots w_{iM}]^T$, associated to hidden state s_i.
d	Dimension of a multivariate Gaussian function.
μ_{ijl}	The lth component regarding the center location of the jth multivariate Gaussian function associated to the hidden state s_i.
$\boldsymbol{\mu}_{ij}$	Center vector of the jth multivariate Gaussian function associated to hidden state s_i. That is, $\boldsymbol{\mu}_{ij} = [\mu_{ij1} \cdots \mu_{ijd}]^T$.
$\mathbf{U}_i$	Set of the center vector associated to the Gaussian mixture function i. That is, $\mathbf{U}_i = \{\boldsymbol{\mu}_{i1} \cdots \boldsymbol{\mu}_{iM}\}$.

$\sigma_{ij}(a, b)$	Covariance between a and b for the jth Gaussian function associated to hidden state s_i.
Σ_{ij}	Covariance matrix of the jth Gaussian function associated to hidden state s_i with structure

$$\Sigma_{ij} = \begin{bmatrix} \sigma_{ij}^2(1, 1) & \cdots & \sigma_{ij}^2(1, d) \\ \vdots & \ddots & \vdots \\ \sigma_{ij}^2(d, 1) & \cdots & \sigma_{ij}^2(d, d) \end{bmatrix}$$

V_i	Set of all covariance matrices associated to the hidden state s_i.
$\mathcal{G}(\mathbf{x}, \mu, \Sigma)$	Multivariate Gaussian function expressed as:

$$\mathcal{G}(\mathbf{x}, \mu, \Sigma) = \exp\left(-(\mathbf{x} - \mu)^T \Sigma^{-1}(\mathbf{x} - \mu)\right)$$

$\phi_i(\mathbf{x})$	Gaussian mixture function associated to the hidden state s_i. In particular, $\phi_i(\mathbf{x}) = \phi(\mathbf{x}, \mathbf{w}_i, \mathbf{U}_i, \mathbf{S}_i)$ which, in turn, is given by:

$$\phi_i(\mathbf{x}) = \sum_{j=1}^{M} w_{ij} \cdot \mathcal{G}(\mathbf{x}, \mu_{ij}, \Sigma_{ij})$$

α_i	In the context of autoregressive Markov models, refers to the AR model coefficients associated to hidden state s_i.
$\mathcal{E}(\cdot)$	Statistical expectation operator.
λ	d-dimensional random variable $\lambda = [x_1 \cdots x_d]^T$.
$\mathcal{G}(\lambda, \mu, \Sigma)$	Multivariate Gaussian function parameterized with mean μ and variance Σ.
$\mathbf{h}_k$	Regressor vector at time instant k.
$\mathbf{H}$	Regression vectors matrix defined as $\mathbf{H} = [\mathbf{h}_1 \cdots \mathbf{h}_N]^T$.

1 Introduction

"We ought to regard the present state of the universe as the effect of its antecedent state and as the cause of the state that is to follow. An intelligence knowing all the forces acting in nature at a given instant, as well as the momentary positions of all things in the universe, would be able to comprehend in one single formula the motions of the largest bodies as well as the lightest atoms in the world, provided that its intellect were sufficiently powerful to subject all data to analysis; to it nothing would be uncertain, the future as well as the past would be present to its eyes."

–Pierre Simon de Laplace

This introductory chapter aims to provide a broad idea behind the problem that will be addressed in the book. As the above quote suggests, reality is too complex to be grasped in its multitude. Hence, the best we can do is to account only for the components that have the largest impact on the reality we are trying to explain. That is, the model we create for explaining a given phenomenon is just a simplification of the reality. The set of all remain interactions, which are neglected during our approximation, will appear as unpredictable components. Even if something is unpredictable, it does not mean that it's fully unknown to the observer. Even randomness has frequently an underlying structure which can be unveiled by several distinct approaches such as hidden Markov models.

1.1 SYSTEM MODELS

"Being able to guess" is a concept that is conveyed by the Greek word *stokhastikos* from which the word stochastic has born. The behaviour of any process can invariably be divided into two components: one that follows exactly the model (mental, mathematical, or of any other nature) that describes its action and other, which is unpredictable, and fails from following it. The first component is the deterministic part of the event and the second the stochastic one. In case of total absence of determinism, the process is said to be random and, by definition, its outcome cannot be predicted by means of any past information.

In practice, it is impossible to have a model that is able to capture all the details of an undergoing process. Deviations between the model output and effective observations will always be present.

Often, dynamical processes in continuous-time, are modelled using systems of differential equations which can lead to an approach designated by state-space model. The usual state-space model formulation regards the deterministic component of a process. The stochastic component can be included by adding a random component whose statistical properties are known or estimated. For example a generic non-linear stochastic state-space formulation of any process can be mathematically represented by (1.1),

$$
\begin{aligned}
\dot{\mathbf{x}}(t) &= F\big(\mathbf{x}(t), \mathbf{u}(t), t\big) + \boldsymbol{\varepsilon}(t) \\
\mathbf{y}(t) &= G\big(\mathbf{x}(t), t\big)
\end{aligned}
\tag{1.1}
$$

where $\mathbf{y}(t) \in \mathbb{R}^p$ is the process output vector, $\mathbf{u}(t) \in \mathbb{R}^m$ the input vector, $\mathbf{x}(t) \in \mathbb{R}^n$ is the system state vector, $(p, m, n) \in \mathbb{N}$ and $\boldsymbol{\varepsilon}(t)$ is a random quantity that represents the measurement error and other phenomenons that are not directly described by the model. The independent variable t can be associated to time or any other quantity such as space for example. Additionally, $F(\cdot)$ and $G(\cdot)$ are the system and measurement functions respectively.

A continuous-time system is one that has an infinite number of states defined continuously along the independent variable. Notice that, it is also possible to describe discrete-time systems using the same approach. For those types of systems, the set of differential equations is replaced by difference equations with the format presented in (1.2).

$$
\begin{aligned}
\dot{\mathbf{x}}(k+1) &= F\big(\mathbf{x}(k), \mathbf{u}(k), k\big) + \boldsymbol{\varepsilon}(k) \\
\mathbf{y}(k) &= G\big(\mathbf{x}(k), k\big)
\end{aligned}
\tag{1.2}
$$

In this type of system description, there is an infinite number of possible states. However, they are only defined at discrete time instants kT_s where $k \in \mathbb{N}$ and $T_s \in \mathbb{R}^+$ is the so called sampling period.

It is important to notice that many systems have a finite number of states. For example, digital systems are particular cases of discrete-time systems with finite number of states. In abstract, the behaviour of all automation systems can be described by finite state machines. As the

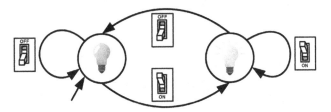

Figure 1.1 A finite state machine representing the behaviour of a bulb lamp as a function of a switch position.

name suggests, a finite state machine is a mathematical description of a class of systems that can be, in a given time instant, in one, and only one state. Figure 1.1 provides an example of a finite state machine associated to the behaviour of a simple domestic lamp bulb as a function of an electric switch position. This graphical way to represent the system behaviour is designated by state diagram [28]. Any state diagram is represented using two types of primitives: circles and arrows.

Circles are used to represent the states that the system can exhibit, and the arrows, the way the system navigates through these states. For example, in Figure 1.1, the system has two states and it is assumed that the initial state is off. It will remain at this state until someone presses the switch. When this happens, the system outcome changes leading to the second state where the lamp is lighted on. Once again, the lamp will remain at this state until the moment that someone presses the switch again and the lamp returns to the off state.

Even if this state diagram is a valid way to represent the relationship between the lamp state and the switch position, no information is provided about when and how this happens. Let's consider that we have monitored a particular lamp status inside a house during some time and have grasped its behaviour. Let's assume there is no deterministic time at which the lamp is lighted and, therefore, only its stochastic nature was captured. After a large set of experiments it has been possible to conclude that, in about 75% of times, if the lamp is on at the present instant, it is expected to remain on from this point onward. On the other hand, if it is off then, in 95% of the observed cases, it has the tendency to remain off. This insight about the lamp behaviour can be represented using the same state machine diagram approach as the one considered earlier. However, now instead of the switch position, the state transitions are associated to probabilities. Figure 1.2 puts forward this concept.

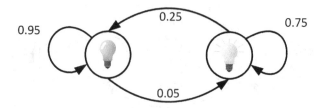

Figure 1.2 A finite state machine representing the behaviour of a bulb lamp as a function of state transition probabilities.

Notice that, since the lamp must always be in one of the two possible states, the sum of the state transitions probabilities is one. This imposition is commonly designated by stochastic constraint. The probability associated to any transition describes how likely it is that this transition will happen. In other words, if the lamp is on, then there is a 75% probability that it will stay on and 25% probability that will go off. The same mental exercise can be made assuming the lamp is off. It is important to notice that the next lamp state only depends on its present state and is independent on its previous history. Probabilistic based systems with such independency between the present and the past are called Markov processes and the model represented in Figure 1.2 a Markov chain.

1.2 MARKOV CHAINS

Before unveiling the concept of Markov chains, let's define what a Markov process is. Markov processes are among the most important of all stochastic processes[1] and are characterized by the fact that their future behavior is statistically independent on their past history. In Chapter 2, the formal meaning of statistical independence will be introduced. Meanwhile, it is sufficient to give the idea that, in this kind of processes, it is assumed that its future outcome does not depend on the past.

A Markov chain is fundamentally related to the concept of Markov processes. Even if many distinct definitions can be provided, in this book a Markov chain is a Markov process with a discrete, and numerable, state space. That is, the number of possible states a given process can have

---

[1]A stochastic process is a sequence of events in which the outcome, at any stage, depends on some probability.

is finite. The hopping between those states is accomplished according to some probability value. Thus, and unlike deterministic systems where from the knowledge of the initial state it's possible to predict exactly the next state, in a Markov chain there is a set of possible states that are probable to occur from the present actual one.

A change from the present state to another from the available state space is known as a transition and, in the context of Markov chains, will occur with a given probability. Within this formulation, it is assumed that the value of those probabilities is static and does not change during the model operation time window.

Formally, a Markov chain can be described by means of two structures: the state transition probability matrix, $\mathbf{A}$, and the vector of initial probabilities $\mathbf{c}$.

For example, assuming that the probability of the lamp to be initially off is 90% and considering that the stochasticity constraint must hold, the Markov chain of Figure 1.2 can be described by:

$$\mathbf{A} = \begin{bmatrix} 0.95 & 0.05 \\ 0.25 & 0.75 \end{bmatrix} \tag{1.3}$$

$$\mathbf{c} = \begin{bmatrix} 0.9 \\ 0.1 \end{bmatrix} \tag{1.4}$$

The diagonal values of the $\mathbf{A}$ matrix represent the probability that the next state remains the same as the present one and the off diagonal values concern the probability of switching to different states. In the same line of thought, the first element of vector $\mathbf{c}$ represents the probability to have the lamp off as the initial lamp state and the second value, the probability to have it on.

Frequently, it is not possible to directly observe the state at which a given process is. Instead, we are only able to observe a set of its outcomes that can be used to infer the underlying statistical description. In those cases, the best model for the process is one that has two distinct layers: one that is responsible to capture the states transitions and the other, correlated to the first, whose job is to generate the observations by mean of probabilistic models. This alternative stochastic model is known as "hidden Markov" model and is the fundamental subject of this book.

1.3 BOOK OUTLINE

Currently, some books that deal with the concept of hidden Markov models are commercially available. Typically, these books are directed to a specific audience in the area of mathematical statistics. Hence, the concepts are usually provided within a deep and formal theoretical framework. This type of approach is harder to follow by readers with an engineering or computer science background. Consequently, this book provides the concept of hidden Markov model in a more "engineer" driven format. That is, some of the mathematical formalisms are relaxed and most of the relevant algorithms are coded in one of the most ubiquitous numerical computation language at the present time: MATLAB®. It is believed that this approach will close the gap between abstract knowledge and objective implementation. Moreover, examples are presented in order to unfold some of the current capabilities and applications of hidden Markov models.

To attain this objective, the present book is structured into six chapters. Chapter 2 presents a brief overview on probability theory. It is not, by any means, an exhaustive or theoretical profound chapter on probability theory. Instead, it is provided here to make this book as self-contained as possible.

Chapter 3 deals with discrete Markov models and presents, in both mathematical and computer format, the three fundamental algorithms associated to discrete observations hidden Markov models: the forward, the backward and the Viterbi algorithms. Their integration into the Baum-Welch parameter estimation procedure is also addressed.

Chapter 4 presents the continuous observations counterpart of the hidden Markov models. As in the previous chapter, the fundamental algorithms related to this model paradigm are presented in both mathematical notation and computer coded.

The hidden Markov models addressed in the two previous chapters are not conceived to deal explicitly with time. This condition is circumvented by means of a third type of approach: the autoregressive Markov model. Its formulation and implementation is addressed at Chapter 5.

Chapter 6 is devoted to present a set of examples where hidden Markov models can be applied. Biomedical and meteorological applications are provided with the algorithms, derived along the previous chapters of the book, applied to them.

2 Probability Theory and Stochastic Processes

"The probability of life originating from accident is comparable to the probability of the unabridged dictionary resulting from an explosion in a printing shop."

–Edwin Grant Conklin

Probability theory is at the heart of hidden Markov models. Its understanding is fundamental to follow the concepts and equations that will be presented in the following chapters. Even if this subject is outside the scope of this book, a brief overview is provided here to make this book as self-contained as possible. During this chapter, the basics on probability theory will be provided without subjecting the reader to all the formalisms usually accompanying this theme.

2.1 INTRODUCTION

The dream of many is to win the lottery. Taking a gamble in the lottery usually relies on selecting a key combination of numbers from an allowable set. For example, choosing six numbers from the universe of integers between one and forty nine.

The choice of those six numbers, which can result in a life changing event, can be done in a myriad of very different ways according to the players character and beliefs: usually an informal selection based on important birthday dates, dreamed numbers, "gut feelings", or many other different superstitions.

While not in the habit of gambling, we the authors decided to take our chances and try it. However, instead of using any of the above enumerated techniques, we have thought that our odds would be substantially improved if we embraced a more "scientific" approach to choose the wining combination.

Thus, the results concerning the draws of the last fourteen years or so, were gathered. It was not a difficult task since this type of data can readily be found and downloaded from the internet. Using this information, the plot illustrated in Figure 2.1 was produced. The horizontal axis presents all the numbers that can be drawn and the vertical axis, the relative frequency of each number occurrence. This kind of plot is designated by histogram. Each one of the histogram bars represents the number of times that a particular number was extracted, during the considered time interval, relatively to the total number of game weeks. That is, if n_y represents the number of times that, in N weeks, the number y is extracted then its relative frequency ν_y is given by:

$$\nu_y = \frac{n_y}{N} \tag{2.1}$$

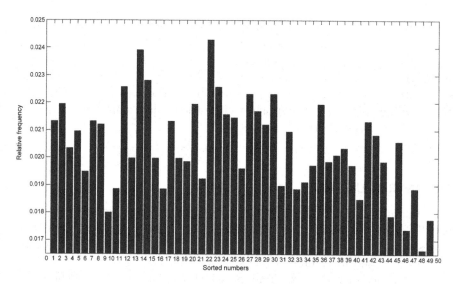

Figure 2.1 Histogram, relative to the extraction of Lottery numbers, since March 1997 up to February 2011 (*Source*: Jogos Santa Casa).

According to the gathered information, a set of different number combinations were produced. For instance, a key that involved the six most frequent numbers, other key that employed the six least frequent numbers, etc. At the end of the process, a total of ten different keys were submitted. After the official extraction, and by comparing the ten bets with the official result, a total disappointment settle down: not even a single number was correct... *so what was wrong in our approach?*

The problem was that we tried to infer the behaviour of the lottery numbers generated system through a sample too small. That is, we tried to catch the nature of a stochastic system having only access to a limited set of observations of that system[1] which, by evaluating the final outcome of our experiment, proves not to be one and the same thing...

In the case of the lottery extraction system, the process is too complex and can not be properly described by a deterministic set of equations. Moreover, even if this was possible, the player, acting as an observer, does not have access to any of the fundamental physical variables. For example, among many others, the momentum of each ball, its initial position, friction, etc. Thus, the unreachable nature of this process, leads us to consider it as a random process. The extraction method is designed in such a way as to deliver an equal probability of occurrence for each number. For example, considering the universe of 49 distinct numbers, the probability of getting one of those numbers during, the first number extraction, is equal to $\frac{1}{49}$. Notice that, the set of those 49 numbers is designated, in probability theory, by sample space. In this case, since all the numbers have an identical probability of occurrence, we can say that the probability distribution is uniform. That is, none of the numbers has a different likelihood of arising and the probability of occurrence is equally diluted along the set of possible numbers.

During the section that follows, three important concepts are introduced: random experiences, probability and distribution. Additionally, the fundamental pillars of probability theory will be established by presenting their fundamental postulates and properties.

2.2 INTRODUCTION TO PROBABILITY THEORY

In engineering, and in science in general, the empirical approach is essential. Normally, if a certain experiment was repeated under similar conditions, then matching results are expected to occur. However, this is not always the case. Frequently, the repetition of an experiment, under very

---

[1]Indeed, is what the scientific method is all about. First performing an observation, then the hypothesis followed by experimentation and conclusion. However, frequently, we are unable to grasp a full understanding of the target phenomenon only by observing a fraction of its full history.

similar conditions, does not necessarily lead to exactly the same outcome. In practice, there is always some stochastic component in all type of experiments performed. This random component constantly adds uncertainty about the exact outcome of an experiment. Hence, the result can only be predicted with a certain probability. For example, and returning to the case of the lottery game, we can only say that the first drawn number will be 1 with a probability of around 2%. Notice that, during the balls extraction, each drawn number can be viewed as an event.

In probability theory, an event refers to a set of outcomes resulting from a given experiment to which a probability is associated. For example, as already referred, the extraction of a number from the lottery wheel is an example of an event. In abstract, an event is always a subset of the sample space which can include the entire sample space or even the empty set, i.e., an event which cannot happen.

There are two important ways, from which, it is possible to estimate the probability of occurrence of a given event:

a priori **probability:** If an event can occur in n different ways, in a total of N possible, and if all are equally probable, then the event probability is $\frac{n}{N}$;

a posteriori **probability:** If, after n repetitions of an experiment, it is observed the occurrence of a given event p times, then the occurrence probability of that event is $\frac{p}{n}$. In this strategy, the repetition number can be high. Thus, when the number of experiments tends to infinite, the occurrence frequency of a given event tends to its probability value.

In the context of probabilities theory, it is important to explicitly define some terms which were introduced in the above paragraphs. Namely:

Experiment: The word "experiment" refers to some arbitrary action that can lead to a given "result" from a pool of possible results;

Random experiment: An experiment is considered random in nature if it can be repeated indefinitely under the same conditions and exhibit an unpredictable outcome. That is, the individual results are unpredictable but, when considered together, some statistical regularity is observed;

Outcome: As pointed out by the name itself, an outcome refers to the observation generated after the execution of an experiment;

Sample space: The sample space refers to the pool of possible outcomes associated to a given experiment. If the set Ω denotes the sample space of some experiment and if ω is an outcome of that experiment, then $\omega \in \Omega$;

Event: An "event" E is any subset of the sample space associated to some experiment. That is $E \subset \Omega$;

Probability: Probability is a quantitative numerical measure, taken in the interval $[0, 1]$, associated to a given "result" in the context of the sample space. That value varies with the actually defined "event" according to a function denoted by *probability mass function*[2].

The remain of this Chapter is organized in the following manner: Section 2.2.1 introduces the concept of *random variable* as the nuclear object of probability theory. Additionally, the relationship between that class of variables and the statistical event is established. During Section 2.2.2 the occurrence probability of an event, in the context of discrete variables, is defined. The concept of probability mass function[3] is also introduced. Finally, Section 2.2.3 presents a set of axioms and properties related to probability theory.

2.2.1 EVENTS AND RANDOM VARIABLES

In the previous section, a statistical event was defined as the set of results from an experiment. Formally, any subset of the sample space is an event. Moreover, any event that consists of only a single sample space result, is denominated by simple event. Events involving more than one result are called composite events. For example, if a digital system sends a three-bit sequence then, from the receiver perspective, the sample space is $\Omega = \{000, 001, 010, 011, 100, 101, 110, 111\}$. Within this sample space, several distinct subsets can be defined and each of those subsets are events. For example, the condition "all sent bits are 0" refers to a simple event given by $E = \{000\}$. On the other hand, the event, "at least one of the transmitted bits is 1", is defined by the composite event $E = \{001, 010, 011, 100, 101, 110, 111\}$.

Assuming a sample space $\Omega = \{\omega_1, \cdots, \omega_N\}$, the set E is considered an event if $E \subset \Omega$. Figure 2.2 illustrates this situation for three distinct events E_1, E_2 and E_3 partially overlapping.

---

[2]This function is related only with discrete random variables. In the case of continuous variables, the concept between probability and probability density function is different.

[3]Usually also called discrete probability density function.

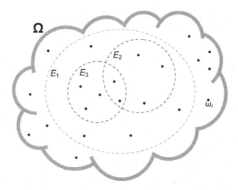

Figure 2.2 Partition of the sample space Ω with N elements, from ω_1 to ω_N, in three subsets E_1, E_2 and E_3.

Often, an event or sample space is associated to a variable, with codomain in $\mathbb{R}$, designated by random variable. This variable is an application that performs the mapping between a point in the sample space Ω into an alternative space. That is, if a random experiment, expressed in Ω space, is considered and if ω_i is any element belonging to Ω, then the random variable $X(\omega_i) \in \mathbb{R}$ is defined[4] as being a function that assigns a real value $\{x_i \in \mathbb{R} : X(\omega_i) = x_i\}$ to each sample $\omega_i \in \Omega$. On other words, it performs the mapping of the sample space into the real line. Figure 2.3 illustrates this concept.

If an event is a subset of the sample space and if, to each individual element of the sample space, one can assign a real value, then the event mapping by the random variable X will be a sub-interval of the codomain of that random variable. To make it clearer, imagine the following universe of persons standing in a room:

$$\Omega = \{\text{John}, \text{Lucy}, \text{Andrew}, \text{Mary}, \text{Ann}\} \tag{2.2}$$

Assume also that the heights and gender of each individual were recorded. The random variable $X(\omega_i)$ is defined as being the "person's height ω_i". For the current sample space, the value of X, expressed in centimeters, is:

$$X(\Omega) = \{168, 161, 175, 155, 162\} \tag{2.3}$$

---

[4]Notice that often, for simplicity of notation, the random variable $X(\omega)$ is compactly described by using only the letter X.

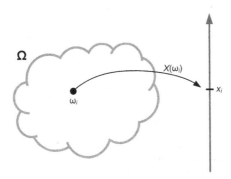

Figure 2.3 The random variable $X(\omega_i)$ viewed as a function that transforms any $\omega_i \in \Omega$ in $x_i \in \mathbb{R}$.

If the event E is now described as being "female with height greater than 160 cm". The subset is now:

$$E = \{\text{Lucy, Ann}\} \tag{2.4}$$

The values of the random variable X associated to this new event are:

$$X(E) = \{161, 162\} \tag{2.5}$$

2.2.1.1 Types of variables

In general, the variables in any problem, can be distinguished as belonging to one of two types: continuous or discrete. A variable is continuous, over an arbitrary interval, if it can take any value within that interval. For example if x represents the voltage between the two terminals of an AC power supply outlet, then it may take any value between −325 V and +325 V relatively to ground[5]. On the other hand, a discrete variable is one that can only take a finite number of values between a given finite range. For instance, the number of bits stored in a *pen drive* of 4 GiB is discrete in nature. In this case, the variable x can only take a finite number of values[6]. For example an integer value between 0 and 2^{35}.

---

[5]The single phase outlet nominal RMS voltage in the EU is equal to 230 V to which corresponds an absolute peak voltage equal to $230\sqrt{2} \approx 325.3$ V.

[6]Without paying any attention to space partition, and just by considering the number of storing bits available.

Both continuous and discrete variables do not need to be constrained to a finite interval. For example, the cumulative number of bits transferred in a data communication network has no upper bound.

Having defined the concept of event and random variable, explaining the concept of probability and probability mass function will be the aim of the next section.

2.2.2 PROBABILITY DEFINITION

There are several ways to introduce the concept of probability. From a classical perspective, the occurrence probability of an event $E \subset \Omega$, characterized by a finite numerable set of elements, is called $P(E)$ and is equal to the ratio between the cardinal[7] of E to the cardinal of Ω. That is,

$$P(E) = \frac{\#E}{\#\Omega} \tag{2.6}$$

If the sets E and Ω are completely described, the value of the probability $P(E)$ is computed *a priori* without requiring the execution of any experience. Furthermore, the probability of an event E is also supported by the following properties:

- The probability of an event E is a real number bounded by:

$$0 \leq P(E) \leq 1$$

  for $E \subset \Omega$.

  Notice that, if $P(E) = 1$, then the occurrence of E is guaranteed. That is, the occurrence probability of E, as result of an experiment with space Ω, is 100%. In the same way, an event E is impossible to happen if $P(E) = 0$.

- Defining a set of N disjoint partitions from Ω, $E_1, \cdots, E_N$, such that $\Omega = E_1 \vee \cdots \vee E_N$, then:

$$\sum_{i=1}^{N} P(E_i) = 1 \tag{2.7}$$

---

[7]The cardinality of a set is a measure regarding the number of elements of that set.

As mentioned earlier in this chapter, the probability can also be evaluated *a posteriori* by analyzing the relative frequency of an event. This is done after performing a sufficiently large number of experiments. When the number of experiments tends to infinity, the relative frequency of the event tends to the value of its occurrence probability. This phenomenon is known as the "law of large numbers" and can be put as follows:

$$P(E) = \lim_{N \to \infty} \nu_N(E) \qquad (2.8)$$

where

$$\nu_N(E) = \frac{n_E}{N} \qquad (2.9)$$

refers to the relative frequency of occurrence of event E in N experiments and n_E represents the number of times that the event E occurred in N experiments. It is worth to note that, as previously suggested, our main reasoning fault in the lottery game was to try to obtain the *a posteriori* occurrence probability without the sufficient number of experiments.

In order to illustrate both definitions of probability, imagine that it is proposed to determine the probability of observing two consecutive identical bits during a three-bit digital transmission[8]. According to expression (2.6), the value of this probability can be calculated as the ratio between the cardinal of $E = \{001, 011, 100, 110\}$ and the cardinal of the sample space $\Omega = \{000, 001, 010, 011, 100, 101, 110, 111\}$. The former has cardinality equal to 4 and the latter equal to 8. Thus, in this case,

$$P(E) = \frac{\#E}{\#\Omega} = \frac{4}{8} = 0.5 \qquad (2.10)$$

Now, suppose that a hypothetical observer, placed downstream the transmission channel, does not know Ω. This observer also aims to obtain the probability of event E but he can't resort to (2.10). Hence, the observer begins to collect and analyse the received data. To each new reception, the observer adjusts the probability value of occurring E. This computation is made by applying expression (2.8) to the incoming observations. Figure 2.4 shows a possible evolution of the estimated probability value, obtained by the observer, as a function of the number of experiments. Listing 2.1 presents the sequence of MATLAB® commands that, after execution, will lead to the above mentioned figure.

---

[8]Assume that the probability of transmitting a '0' is equal to the probability of transmitting a '1'.

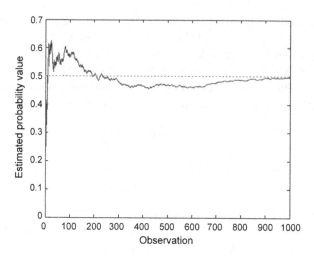

Figure 2.4 Evolution of the estimated probability value of "receiving two consecutive identical bits in a three bit word" as a function of the number of retrieved observations.

Listing 2.1 *A posteriori* probability estimation of getting two identical bits in a three bits transmission.

```
1  clear all;
2  clc;
3  N=1000; %------------------------Number of observations
4  for k=1:N,
5      B=rand(1,3)>0.5;%-------------Transmits a 3 bit word
6      E(k)=abs(sum(diff(B)));  %----1 if there are only two equal bits
7                               %    in a row and 0 otherwise
8      P(k)=sum(E)/k;
9  end
10 plot(1:N,P,1:N,0.5*ones(N,1),'--');
11 axis([1 1000 0 1]);
12 xlabel('Observation');
13 ylabel('Estimated probability value');
```

As can be seen, the fluctuation in the probability estimation is decreasing as the number of observations increase and is gradually converging to the actual value of the probability which, as seen earlier, is equal to 50%.

A mathematical formulation, related to the previous definition of probability, can be considered. In the case of discrete random variables, this function is denoted by probability mass function and can be formally described by:

$$f_X(x) : x \in X, \ f_X(x) = P(X = x) \tag{2.11}$$

16

Notice that, for random discrete variables, the relationship between the probability mass function for a given value x belonging to X, and the probability of X being equal to x is the same. Moreover, it is important to underline that the function $f_X(x)$ is defined for any x. Even for those values that do not have correspondence into the sample space set. In this case, $f_X(x) = 0$ if $x \notin X$.

The concept of probability mass function can be conveniently illustrated by revisiting the problem of the three-bit digital transmission system. It has already been said that the sample space is given by:

$$\Omega = \{\omega_0, \omega_1, \omega_2, \omega_3, \omega_4, \omega_5, \omega_6, \omega_7\} \tag{2.12}$$

where $\omega_0 = 000$, $\omega_1 = 001$, $\omega_2 = 010$, $\omega_3 = 011$, $\omega_4 = 100$, $\omega_5 = 101$, $\omega_6 = 110$ and $\omega_7 = 111$.

Consider the event E, defined verbally as "exactly two consecutive bits are identical", which corresponds to the set $E = \{\omega_1, \omega_3, \omega_4, \omega_6\}$.

The random variable $X(\omega)$ is now defined as being "the number of identical consecutive bit pairs in a sent message". In this case, the codomain of X is the set $\{0, 1, 2\}$. For example $X(\omega_0) = 2$, $X(\omega_1) = 1$ and $X(\omega_2) = 0$. The mapping of all the values of the sample space leads to the following set:

$$X(\Omega) = \{2, 1, 0, 1, 1, 0, 1, 2\} \tag{2.13}$$

Figure 2.5 illustrates this mapping for some selected elements of Ω. In this problem, it is assumed that every observation ω_i in Ω is equally likely to occur. That is,

$$P(\omega_i) = \frac{1}{8}, \quad i = 0, \cdots, 7 \tag{2.14}$$

It has already been seen that the occurrence probability of event E is equal to:

$$P(E) = \frac{1}{2} \tag{2.15}$$

In this framework, the probability mass function is defined as follows:

$$f_X(x_i) = \begin{cases} \frac{2}{8} & \text{for } x_i = 0 \\ \frac{4}{8} & \text{for } x_i = 1 \\ \frac{2}{8} & \text{for } x_i = 2 \end{cases} \tag{2.16}$$

where the values of the probabilities $P(X = x_i)$, for each of the three possible values of x_i, are obtained by applying (2.9).

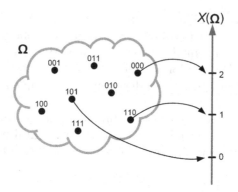

Figure 2.5 Example of mapping some elements of the sample space Ω in the real line $X(\Omega)$.

In this case, the value of $P(E)$ equals $P(X = 1)$. Hence, the probability value of occurrence of event E is equal to the probability of one pair of identical bits being transmitted in the message. That is,

$$P(E) = f_X(1) \tag{2.17}$$

More generically, the probability of an event E, defined by N elements $\{\omega_1, \cdots, \omega_N\}$, is related to the probability mass function through:

$$P(E) = \sum_{x_i=X(\omega_i)} f_X(x_i), \quad i = 1, \cdots, N \tag{2.18}$$

For example, if the random variable X is now described as the "value, expressed in decimal of the modulo-2 sum of the incoming message bits"[9], then:

$$X(\Omega) = \{0, 1, 3, 2, 2, 3, 1, 0\} \tag{2.19}$$

The probability mass function is now:

$$f_X(x_i) = \begin{cases} \frac{2}{8} & \text{for } x_i = 0 \\ \frac{2}{8} & \text{for } x_i = 1 \\ \frac{2}{8} & \text{for } x_i = 2 \\ \frac{2}{8} & \text{for } x_i = 3 \end{cases} \tag{2.20}$$

---

[9]Assuming a *big endian* transmission format where the bits are sent in the order of greatest significance.

The occurrence probability of E is equal to the probability of X being equal to 1 plus the probability of X being equal to 2 that is[10],

$$P(E) = \sum_{x_i=1, x_1=2} f_X(x_i) = \frac{2}{8} + \frac{2}{8} = \frac{1}{2} \qquad (2.21)$$

This example also shows that, there is always some subjectivity in the choice of the random variable for a given problem. In other words, the same problem can be solved by defining different formats for a random variable.

The probability mass function can often be represented graphically. For example, both functions (2.16) and (2.20) can be plotted as illustrated in Figure 2.6.

In addition to the probability density function, other notions must be introduced in order to fully grasp the concepts which will be provided in the upcoming chapters. In particular, the fundamental axioms and properties of probability theory. Hence, the following section will be devoted to introduce them.

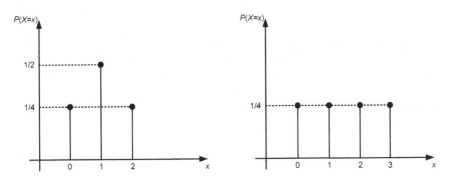

Figure 2.6 Graphical representation of the probability mass function. On the left the aspect of the function (2.16) and on the right of the function (2.20).

2.2.3 AXIOMS AND PROPERTIES

The occurrence probability of a given event A is denoted by $P(A)$ and must verify the following axioms:

---

[10]This can be easily inferred by comparing the set (2.13) with (2.19).

- The probability of occurrence of an event A is always constrained to the interval:

$$0 \leq P(A) \leq 1 \qquad (2.22)$$

- The probability of an event A not to occur is given by:

$$P(\overline{A}) = 1 - P(A) \qquad (2.23)$$

Additionally, the formal definition of probability requires that:

- The probability of an impossible event is zero:

$$P(\varnothing) = 0 \qquad (2.24)$$

- The probability of a certain event is one:

$$P(\Omega) = 1 \qquad (2.25)$$

The probability $P(A)$ referred above, is frequently known as "marginal probability" when other types of probabilities are at stake such as "joint probability" and "conditional probability".

If a second event, say B, is also considered, the joint probability between A and B, refers to the probability of simultaneous occurrence of both events and is denoted as $P(A \wedge B)$. The concept of joint probability leads to one of the most fundamental equalities known as the "law of total probability" which can be expressed as:

$$P(A) = P(A \wedge B) + P(A \wedge \bar{B}) \qquad (2.26)$$

Another important notion is the one of conditional probability. As the name may suggest, this probability concerns the one that derives from the probability of observing an event A, constrained to the occurrence of a second event B. This type of probability is expressed as $P(A|B)$.

All the three above referred types of probabilities are connected by one of the most fundamental equalities in probability theory: Bayes theorem. This theorem can be written as:

$$P(A|B) = \frac{P(A \wedge B)}{P(B)} \qquad (2.27)$$

where, by convention, if $P(B) = 0$ then $P(A|B) = 0$.

Notice that the Bayes theorem can appear in different shapes. For example, it can be formulated as:

$$P(B|A)P(A) = P(A|B)P(B) \qquad (2.28)$$

or, in the case of three different events A, B and C, as:

$$P(A|B \wedge C) = \frac{P(A|C)P(B|A \wedge C)}{P(B|C)} \qquad (2.29)$$

It is straightforward to demonstrate the above equality by resorting to (2.2) and taking into consideration the following relation:

$$\begin{aligned} P(A \wedge B \wedge C) &= P(A)P(B|A)P(C|A \wedge B) \\ &= P(C)P(B|C)P(A|B \wedge C) \end{aligned} \qquad (2.30)$$

usually denoted as the "chain rule".

A concept closely related to the Bayes rule is that of statistical independence between two events. An event A is statistically independent of a second event B if[11],

$$P(A|B) = P(A) \qquad (2.31)$$

If this is the case, then the joint probability between A and B can be computed by the product of the marginal probabilities of A and B. That is,

$$P(A \wedge B) = P(A) \cdot P(B) \qquad (2.32)$$

The concept of conditional probability can be extrapolated to the cases where more than one variable is involved. For example, the conditional probability of event A given B and C is described as $P(A|B \wedge C)$. In this context, the concept of statistical independence can also be extended to the case of more than two variables. Thus, if $P(A|B \wedge C)$ does not depend on the occurrence of B,

$$P(A|B \wedge C) = P(A|C) \qquad (2.33)$$

which means that A is "conditionally independent" of B given C. From this definition, it is possible to establish the following equalities:

---

[11]Some literature uses the notation $A \perp B$ to indicate that A is statistically independent of B.

$$P(A \wedge B | C) = P(A | C) \cdot P(B | C) \qquad (2.34)$$

$$P(B | A \wedge C) = P(B | C) \qquad (2.35)$$

Before ending this section, it remains to consider the probability of the union between two events. Assuming again two events A and B, then the probability of the union between those two events, denoted as $P(A \vee B)$, is equal to:

$$P(A \vee B) = P(A) + P(B) - P(A \wedge B) \qquad (2.36)$$

Remark that, if A and B are "mutually exclusive" events, that is, the occurrence of one prevents the occurrence of the other, then $P(A \wedge B) = 0$ and the above expression can be rewritten as:

$$P(A \vee B) = P(A) + P(B) \qquad (2.37)$$

As a final consideration, it is worth noticing that (2.36) can also be extended to handle an arbitrary number of events. For example, and for the particular case of three events A, B and C, the probability of the union of those three events can be computed by:

$$P(A \vee B \vee C) = P(A) + P(B) + P(C) - P(A \wedge B) - P(A \wedge C)$$
$$- P(B \wedge C) + P(A \wedge B \wedge C)$$

$$(2.38)$$

2.3 PROBABILITY DENSITY FUNCTION

Up to now, the concept of probability was derived assuming a finite number of outcomes from a given experiment. However, in many situations, this is not the case. Frequently, there is an infinite number of different values that an event can take. For example, when the random variable X is continuous. In those cases, we cannot talk about probability distribution function. Instead, the concept of probability density function must be introduced.

In probability theory, a probability density function is a function that provides the value of the relative probability of a continuous random variable X to be equal to some value taken from the sample space.

It is fundamental to notice that the absolute probability of a random variable to take any value within the sample space is equal to zero. This statement can be demonstrated by considering the following example. If one is sending a binary message with n bits, what is the probability of observing all the bits equal to zero if one assumes equal probability of observing either a '1' or a '0'? The answer is simple and leads to $\frac{1}{n}$. For example, assuming an one bit word, the probability is equal to 50%. If one byte is considered, then the probability decreases to near 0.4 % and for 1 gigabyte then it is around 8.3×10^{-23} %. So, we are expecting to have a probability of occurrence that will tend to zero when we increase the number of bits toward infinity. Hence, using the probability distribution function, it is possible to compute the probability of the continuous random variable to be within a given interval as opposed to take a precise value.

In order to $p(x)$ be considered a probability density function, it must exhibit the following two properties:

- $p(x) \geq 0$ for all $x \in X$

- $\int_{-\infty}^{+\infty} p(x)dx = 1$

Assuming that an event A represents the occurrence of a value $x \in X$ greater or equal to some constant a and a second event B the occurrence of a value x lower or equal to b, and if $a \leq b$ then,

$$P(A \wedge B) = \int_{a}^{b} p(x)dx \qquad (2.39)$$

where $p(x)$ is the probability density function associated to X.

In the same line of thought, it is possible to say that,

$$P(B) = \int_{-\infty}^{b} p(x)dx \qquad (2.40)$$

There are many possible different probability density functions. The most widespread is the one described by the Gaussian function:

$$\mathcal{G}(x, \mu, \sigma) = \left(\frac{1}{\sqrt{2\pi\sigma^2}} \right) \exp \left(\frac{(x - \mu)^2}{\sigma^2} \right) \qquad (2.41)$$

where the parameters μ and σ^2 refer to the mean and variance of the probability density function. The concept of both mean and variance will be addressed in the next section.

2.4 STATISTICAL MOMENTS

By definition, the precise outcome of a random experiment is impossible of being anticipated. However, if a sufficiently large number of observations is available, it is possible to establish an estimate for its value. If no *a priori* knowledge on the generating process is known, the value usually attributed to this estimate is the arithmetic mean. For example, assuming that x_1, x_2, $\cdots$, x_N are the outcomes of a set of N random experiments, the mean value of the observations, denoted by $\bar{x}$, is calculated by:

$$\bar{x} = \frac{1}{N} \sum_{i=1}^{N} x_i \tag{2.42}$$

In order to illustrate the mean value, as an estimation tool for the outcome of a random experiment, suppose that a ball is thrown and the distance that it reaches is measured. The results, after 10 trials, are summarized in Table 2.1.

Table 2.1

Distance hit by throwing a ball.

Trial	1	2	3	4	5	6	7	8	9	10
distance/m	10.5	11.8	7.7	10.9	10.3	8.7	9.6	10.3	13.6	12.8

By applying (2.42), the observations average is equal to 10.6 meters. If someone asks us to provide an estimate for the the distance of an additional 11th trial, the most reasonable value to answer would be 10.6 meters.

While in the process of throwing a ball, let's suppose that we switch to a second ball substantially lighter than the first. The results after 10 experiments, are reported in Table 2.2.

Table 2.2

Distance reached by the launch of a second ball.

Trial	1	2	3	4	5	6	7	8	9	10
distance/m	4.5	4.8	4.6	4.7	31	4.4	4.6	38	4.6	4.8

The mean value in this case is identical to that obtained using the values in Table 2.1. However, the tabulated values are not as consistent as those obtained in the previous trials. Indeed, by looking at Table 2.2, two quite different values are immediately identified. Namely the ones from the 5th and 8th tests. The discrepancy between these two values and the remainder could be due to uncontrollable external phenomena. For example, wind blowing at the instant the ball is thrown, could be the reason why the ball has travelled a greater distance.

The important thing to conclude from this example is that usually the mean, on its own, does not tell the full story. Hence, the average value should always be accompanied by a complementary measure that reflects the confidence in the average value as an estimate. This measure is called variance and is calculated as the mean value of the quadratic difference between the set of experimental results and its mean. Algebraically it is described as:

$$\sigma^2 = \frac{1}{N} \sum_{i=1}^{N} (x_i - \bar{x})^2 \tag{2.43}$$

A measure related to variance is the standard deviation being calculated by:

$$\sigma = \sqrt{\frac{1}{N} \sum_{i=1}^{n} (x_i - \bar{x})^2} \tag{2.44}$$

Both provide the same insight about the process. However, the standard deviation has the advantage of leaving the units with which the value of the random variable is recorded. For example, the standard deviation and variance, associated to the results compiled in Tables 2.1 and 2.2, are presented in Table 2.3.

Table 2.3

Mean, variance and standard deviation of the values presented in the Tables 2.1 e 2.2.

	Mean/m	Variance/m^2	Standard deviation/m
Table 2.1	10.6	3.18	1.78
Table 2.2	10.6	161	12.7

As can be seen, although the mean values are identical, there is a strong discrepancy between the standard deviation (and variance). The deviation in the first set of tests is much smaller than the one obtained in the second set. This means that the mean value in the first set of tests "fits" into the observations with an average error[12] of around 1.8 meters. The uncertainty in the second set of data is substantially greater as can be concluded from its standard deviation value.

The average value in the second set of tests is not as representative of what actually happens when the ball is thrown. This lack of representativeness is mainly due to those two tests whose distances differ substantially from the rest. These trials, although with high values, happen less frequently than the others. Therefore a more representative value could be obtained from a weighted average instead of an arithmetic mean. In particular, greater weight should be associated to the most probable observations and less weight to less probable observations. In fact, this operation is called statistical expectation, denoted by $\mathfrak{E}(X)$, and is mathematically described by[13]:

$$\mathfrak{E}(X) = \sum_{i=1}^{N} f_X(x_i) \cdot x_i \qquad (2.45)$$

where X refers to a random variable such that $X(\Omega) = \{x_1, \cdots, x_N\}$ with probability distribution function $f_X(x_i)$.

Comparing (2.42) with (2.45) it is straightforward to note that the arithmetic mean is nothing more than the expectation of a random variable assuming that all its values have equal probability of occurrence. That is, when

$$f_X(x_i) = \frac{1}{N} \qquad (2.46)$$

The expectation, as an operator over a random variable X, supports the linearity property. So, if $X_1, X_2, \cdots, X_n$ are n random variables and $k_1, \cdots, k_n$ are n constants, then:

$$\mathfrak{E}(k_1 X_1 + k_2 X_2 + \cdots + k_n X_n) = \\ k_1 \mathfrak{E}(X_1) + k_2 \mathfrak{E}(X_2) + \cdots + k_n \mathfrak{E}(X_n) \qquad (2.47)$$

---

[12]Indeed, it is the square root of the quadratic error often referred to as RMS (root-mean square).

[13]This expression is only valid for discrete random variables which are described by probability distribution functions.

Statistical expectation is also termed as the "first statistical moment". In statistics, the moment of order p, of a random variable X with probability distribution $f_X(x)$, is defined by:

$$\mathfrak{E}(X^p) = \sum_{i=1}^{N} x_i^p \cdot f_X(x_i) \tag{2.48}$$

Note that the mean square value of a random variable is nothing more than the second order moment. Moreover, it is also possible to define the central moment of order p of a random variable X as,

$$\mathfrak{E}\left((X - \mathfrak{E}(X))^p\right) = \sum_{i=1}^{N} (x_i - \mathfrak{E}(X))^p \cdot f_X(x_i) \tag{2.49}$$

As can be easily confirmed, if $p = 2$ and if $f_X(x_i) = \frac{1}{N}$, then the second-order central moment acquires the name of variance.

2.5 SUMMARY

The probability theory is a way to discover underlying patterns in random processes. Its roots can be traced, back to more than 500 years ago and, since then, it has become theoretically mature with applications in a broad range of scientific areas. Some of its elementary concepts are lightly addressed in the previous sections. It is important to notice that the current chapter is not intended to provide any profound understanding of probability theory. Instead, its aim is just to provide the basic definitions and concepts that will be required in the next chapters. In particular, the main axioms and properties associated to probability analysis.

The main concepts to retain, after reading this chapter, regard the notions of statistical independence and conditional probability. It is also important to remember how the probability of disjoint events can be computed and how Bayes theorem can be applied to obtain the probability of an event given the prior knowledge on a second event. The next chapter will address the Bayes theorem while introducing the discrete hidden Markov model as a way of modelling dynamic processes from sequences of observed data points.

3 Discrete Hidden Markov Models

"If you can't describe what you are doing as a process, you don't know what you're doing."

–W. Edwards Deming

Hidden Markov models play an important role in dynamic systems modelling. They can be found in a wide range of signal processing applications. In particular, within speech and natural language processing or time-series modelling. This statistical model paradigm can be fully described by means of stochastic equations. However, for actual applications, the conceptual ideas behind this formulation must be translated into a computer program. Since usually the relationship between the mathematical description and the computer codding process is not obvious, in this chapter the hidden Markov model computer implementation will be unveiled by relating the theoretical stochastic equations to a set of functions written in a high-level computer language. Although many other alternatives could be equated, in this book the MATLAB® language was chosen to be the numerical computation tool to be used.

This chapter discusses the three fundamental algorithms associated with the operation of discrete hidden Markov models: the forward, backward and Viterbi algorithms. Their integration into the Baum-Welch training process will also be addressed.

3.1 INTRODUCTION

Hidden Markov models (HMM) were first presented by L. Baum and his co-authors in the late sixties and early seventies of the previous century [2, 4, 3]. Since then, they have been extensively used in various fields of scientific knowledge. Particularly in areas involving signal processing, such as speech recognition [23, 21, 14], biological signal analysis [7] or time-series forecasting [17, 40]. A hidden Markov model has its behaviour described along two distinct dimensions: one denoted by "observable" and the other by "unobservable" or "hidden". In both plans of existence, the basic unit of processing is the "state". States associated to the unobserved plane are called hidden states and, similarly, those in the observable part are designated by observable states. Figure 3.1 shows a schematic representation of the above described idea.

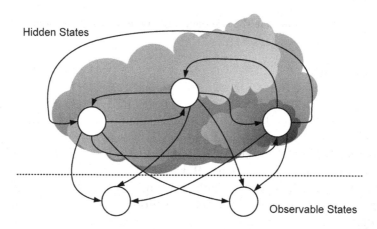

Figure 3.1 The two hidden Markov model layers: the hidden layer, where the states, presented by circles, are interconnected by probabilistic connections and the observable layer, where the observable states are connected to the hidden ones.

The arcs, connecting the states, define the information propagation direction. This information propagation, flowing from one state to the other, happens according to a given probabilistic set of rules. Thus, a hidden Markov model can be understood as a finite state machine where transitions between states are probabilistic. Furthermore, it is fundamental to refer that the hidden dimension satisfies the Markov property. That is, the current active hidden state only depends on the previous active hidden state and is independent of any past model history. On the other hand, the observable dimension depends entirely on the hidden states activity and usually represents the values actually measured during an experiment. This concept can be compared to the case of an ill person. The symptoms of sickness are what someone is able to observe but it is the concealed disease that is actually causing them. In this context, the disease can be viewed as existing in an unreachable hidden dimension while its symptoms, which are what the beholder has access to, as laying at an observing level.

In this book, the hidden states will be designated by the character s followed, in subscript, by the state number. In the same way, observable states will be denoted by r and also followed by a subscript identifying the state number.

In order to be algebraically manipulable, a hidden Markov model is usually assumed to follow the next two conditions:

There are only first-order dependencies: This condition is associated to the hidden states chain and requires that the hidden state q_k, active at a discrete time instant k, only depends on the previous time instant active state. That is, the hidden states long term history is irrelevant to the present state activation. This statement can be expressed as follows:

$$P\left(q_k \mid q_{k-1}, q_{k-2}, \cdots\right) = P\left(q_k \mid q_{k-1}\right) \tag{3.1}$$

where $P(\cdot)$ should be read as the probability of occurrence of the situation described in its argument. In expression (3.1) the argument regards the situation of having a particular active state at time k conditioned to the fact that all the active hidden states history is known. Note that, if the hidden chain has a total of m states, then $q_k \in \{s_1, \cdots, s_m\}$. Hence (3.1) defines that the probability of activating one of the m hidden states, at time k, depends only on the active state at $k - 1$.

There is complete independence between observations: Let us assume a hidden Markov model with m hidden states and n observable states. Consider also that $\lambda_k \in \{r_1, \cdots, r_n\}$ refers to the active observable state at time k. The conditional independence among observations can be written as follows:

$$P\left(\lambda_k \mid \lambda_1 \cdots \lambda_{k-1}, q_1 \cdots q_k\right) = P\left(\lambda_k \mid q_k\right) \tag{3.2}$$

Hence, the probability of observing a particular state at time k, in an universe of n states, given the knowledge of all the active states history for both hidden and observable dimensions, reduces to the probability of having λ_k active conditioned only to the knowledge of the present active hidden state q_k.

It is worth noting that a hidden Markov model is not statistically independent and identically distributed[1]. This is because, in general,

$$P\left(\Lambda_k\right) \neq \prod_{i=1}^{k} P\left(\lambda_i\right) \tag{3.3}$$

where Λ_k refers to the set of all active observations up to the instant k. That is, $\Lambda_k = \lambda_1, \cdots, \lambda_k$. The reason is simple: even if λ_k is only dependent on q_k and λ_{k-1} on q_{k-1}, the present active hidden state q_k depends on q_{k-1}.

Besides these two restrictions, in this book it is assumed that Markov models are both homogeneous and ergodic. A Markov chain is said to be homogenous if the transition probabilities values are not time varying. That is, the transition probability from the hidden state s_i to the hidden state s_j is always the same and does not depend on k. Furthermore, a Markov model is ergodic if, at any instant k, it is possible to move from a hidden state s_i to any of the m states that constitutes the chain.

With these fundamental notions in mind, the next section will be devoted to introduce the concept of state transition matrices. Furthermore, the section will also explain how a Markov chain evolves in time. Special attention is given to the computational load associated with the probabilities calculation. This computational complexity is made evident even for models with a small number of states.

3.2 HIDDEN MARKOV MODEL DYNAMICS

Figure 3.2 represents the time evolution, along N epochs, of a hidden Markov model with m hidden and n observable states. The hidden states are denoted by s_1 to s_m and the observable ones by r_1 to r_n. The transition probability between hidden states i and j is represented by a_{ij}. In the same way, the transition probability between hidden state i and observable state j is b_{ij}. The probabilities c_i for $i = 1, \cdots, m$ are each hidden state startup probability[2]. That is, $c_i = P(q_1 = s_i)$.

---

[1] A set of random variables are said to be independent and identically distributed if all of them have the same probability distribution and are statistically independent.

[2] Usually denoted by *priors*.

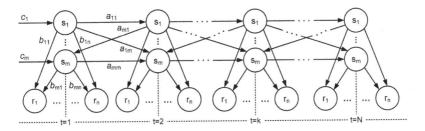

Figure 3.2 Diagram for N epochs time evolution of a hidden Markov model with m hidden and n observable states.

In a generic time instant $t = k$, the activation probability of an arbitrary observable state λ_k is given by the sum:

$$P(\lambda_k) = \sum_{i=1}^{m} P(\lambda_k | q_k = s_i) P(q_k = s_i) \tag{3.4}$$

This means that the probability of observing λ_k is computed from the contribution of each of the possible hidden states. In particular, the above expression represents a weighted sum of m factors of the form $P(q_k = s_i)$. Each of those factors corresponds to the activation likelihood of a specific hidden state.

Any hidden Markov model is fully characterized after defining its probability transition values. Let $\Theta \rightarrow (\mathbf{A}, \mathbf{B}, \mathbf{c})$ represent an ordered tern whose components are the Markov model parameters. In particular, $\mathbf{A}$ is the transition probabilities matrix, $\mathbf{B}$ is the observation probabilities matrix (also referred to as emissions matrix) and $\mathbf{c}$ refers to a vector filled with each hidden state initial probability. The matrix $\mathbf{A}$ is a square matrix of dimension $m \times m$ where each matrix element represents the probability of moving from one state to another. This matrix is formulated as:

$$\mathbf{A} = \begin{bmatrix} a_{11} & \cdots & a_{1m} \\ \vdots & \ddots & \vdots \\ a_{m1} & \cdots & a_{mm} \end{bmatrix} \tag{3.5}$$

where,

$$a_{ij} = P(q_k = s_j | q_{k-1} = s_i) \tag{3.6}$$

refers to the transition probability from the hidden state s_i to the hidden state s_j.

One of this matrix properties is that the sum, along its rows, is always unitary. That is,

$$\mathbf{A} \cdot \mathbf{1}^T = \mathbf{1}^T \tag{3.7}$$

where $\mathbf{1}$ denotes a m dimensional row vector with all its elements equal to one. This stochasticity condition implies that, in the next time epoch, a transition from the current state will surely occur.

The matrix $\mathbf{B}$ describes the transition probabilities between the hidden and observable states. It is a rectangular matrix with dimension $m \times n$ with the following internal structure:

$$\mathbf{B} = \begin{bmatrix} b_{11} & \cdots & b_{1n} \\ \vdots & \ddots & \vdots \\ b_{m1} & \cdots & b_{mn} \end{bmatrix} \tag{3.8}$$

where,

$$b_{ij} = P(\lambda_k = r_j | q_k = s_i) \tag{3.9}$$

denotes the transition probability from the hidden state s_i to the observable state r_j.

Similarly to the $\mathbf{A}$ matrix, the sum along the rows of $\mathbf{B}$ must be equal to one. This means that, at any time instant, one of the n possible output states will surely be observed. This stochastic constraint is formulated in matricial form as:

$$\mathbf{B} \cdot \mathbf{1}^T = \mathbf{1}^T \tag{3.10}$$

where, in this case, $\mathbf{1}$ is a row vector with dimension $1 \times n$.

Finally, $\mathbf{c}$ is a m dimensional row vector whose elements represent the model initial probability distribution. If c_i, for $i = 1, ..., m$, is the initial s_i hidden state activation probability then,

$$\mathbf{c} = \begin{bmatrix} c_1 & \cdots & c_m \end{bmatrix} \tag{3.11}$$

This vector is also bounded to the stochasticity condition. That is,

$$\mathbf{c} \cdot \mathbf{1}^T = 1 \tag{3.12}$$

Now suppose a set of observations $\Lambda_N = \lambda_1 \cdots \lambda_N$ where, as already mentioned, $\lambda_k \in \{r_1, \cdots, r_n\}, \forall k \in \mathbb{N} : 1 \leq k \leq N$. This formulation reflects, without assuming any particular situation, which states are observed at each time epoch. That is, at any discrete time instant k, the observable state can be any one of the n possible states.

The probability of detecting the sequence of N observations Λ_N, from a hidden Markov model with parameters Θ, is $P(\Lambda_N|\Theta)$. Please observe that, due to notation convenience, from now on the explicit dependence between a sequence of observations and Θ will be assumed. Thus, consequently, $P(\Lambda_N|\Theta)$ will be compactly represented by $P(\Lambda_N)$.

According to the chain rule,

$$P(\Lambda_N) = P(\lambda_1 \cdots \lambda_N) = P(\lambda_1) \cdot \prod_{i=2}^{N} P(\lambda_i | \lambda_1 \cdots \lambda_{i-1}) \qquad (3.13)$$

and, since complete independence between observations is assumed, the previous expression takes the form:

$$P(\lambda_1 \cdots \lambda_N) = P(\lambda_1) \cdot P(\lambda_2) \cdot P(\lambda_3) \cdot \ldots \cdot P(\lambda_N) \qquad (3.14)$$

This last equality can be interpreted as follows: the probability of obtaining a given sequence is equal to the product of the occurrence probabilities of each particular observation.

Now, the question is on how to compute the probability $P(\lambda_k)$. That is, the probability of observing λ_k at time k. The response depends on the current active hidden state. If the model is presently at state s_1, the probability of obtaining λ_k is equal to the transition probability from the hidden state s_1 to the observable state λ_k. This quantity is denoted by $P(\lambda_k|q_k = s_1)$. On the other hand, if the machine is on an alternative hidden state, say i, the probability of obtaining λ_k is now $P(\lambda_k|q_k = s_i)$ for $i = 1, \cdots, m$. Thus, the probability of observing state λ_k is equal to the probability of obtaining this state assuming that the machine is in state s_1 or in state s_2 or ... or in state s_m. Hence, the probability of observing λ_k is equal to,

$$\begin{aligned} P(\lambda_k) &= \sum_{i=1}^{m} P(\lambda_k | q_k = s_i) P(q_k = s_i) \\ &= \sum_{i=1}^{m} b_i(\lambda_k) \cdot P(q_k = s_i) \end{aligned} \qquad (3.15)$$

where $P(q_k = s_i)$, for $i = 1 \cdots m$, denotes the probability of having, at time instant k, the hidden state s_i active. Moreover, the newly introduced function $b_i(\lambda_k)$, is equal to $P(\lambda_k|q_k = s_i)$. Hence, if $\lambda_k = r_j$, then $b_i(\lambda_k) = b_{ij}$ where b_{ij} is the ith row and j^{th} column element of the $\mathbf{B}$ matrix.

The computation complexity of the above expression is mainly due to the process of finding the likelihood of attaining, at an arbitrary time instant k, a particular hidden state. At the initial instant $(k = 1)$, the probability of the machine be in an arbitrary hidden state i, to $i = 1, \cdots, m$, is equal to c_i. However, things become more complicated for later time instants. For example, the probability of reaching an arbitrary state s_i at $k = 2$, is the sum of the probabilities of reaching that state by all states upstream of the network represented in Figure 3.2. Figure 3.3 illustrates this by marking all m possible paths that lead to s_i.

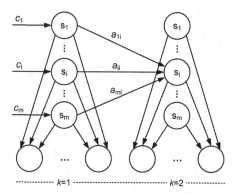

Figure 3.3 Representation of all possible paths that lead to s_i, in $k = 2$, for $i = 1, \cdots, m$.

It is not hard to imagine that the number of possible paths, leading to an arbitrary hidden state, increases exponentially with k. As a matter of fact, the number of different paths that can lead to an arbitrary hidden state, at $t = k$, is equal to m^{k-1}. As an example, for $k = 3$, the number of possible paths to some arbitrary hidden state is equal to m^2. Figure 3.4 illustrates this issue by representing the trellis of possible paths that lead to the hidden state s_i.

Even for a small number of states, finding the probability of reaching a given state, on a more or less distant horizon, becomes prohibitive in terms of computational complexity. However, instead of using a "brute force" approach to this problem, a more computationally efficient method can be used. This alternative method takes advantage of past computed results in order to efficiently obtain new values for each time instant. This recursive strategy is known by "Forward Algorithm" and will be presented in the following section.

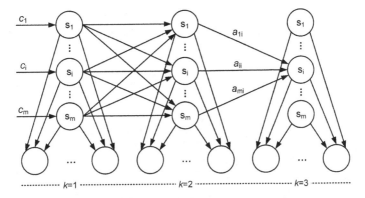

Figure 3.4 Representation of all possible paths that lead to s_i at the parametric instant $k = 3$.

3.2.1 THE FORWARD ALGORITHM

Consider a sequence of N observations, $\Lambda_N = \{\lambda_1, \cdots, \lambda_N\}$. What is the probability of this sequence has been generated by a hidden Markov model with parameters Θ?

The probability will be the sum of the probabilities, for the present observations sequence, given each of the possible hidden states combinations. As an example, for a two hidden state Markov model, as presented in Figure 3.5, there are four possible active hidden states combination history when at $k = 2$: $\{q_1 = s_1, q_2 = s_1\}$, $\{q_1 = s_1, q_2 = s_2\}$, $\{q_1 = s_2, q_2 = s_1\}$ and $\{q_1 = s_2, q_2 = s_2\}$.

For this reason, the probability of getting the observation sequence $\Lambda_2 = \{\lambda_1, \lambda_2\}$ can be computed using the following expression:

$$P(\Lambda_2) = P(\Lambda_2|\,s_1 s_1)\,P(s_1 s_1) + P(\Lambda_2|\,s_1 s_2)\,P(s_1 s_2) + \\ P(\Lambda_2|\,s_2 s_1)\,P(s_2 s_1) + P(\Lambda_2|\,s_2 s_2)\,P(s_2 s_2) \tag{3.16}$$

where,

$$P(\Lambda_2|s_i s_j) = P(\lambda_1|s_1)P(\lambda_2|s_2) \tag{3.17}$$

that is,

$$P(\Lambda_2|\,s_i s_j)\,P(s_i s_j) = \\ P(\lambda_1|\,s_i)\,P(s_i)\,P(\lambda_2|\,s_j)\,P(s_j|\,s_i) \tag{3.18}$$

Now, let's assume a new observation λ_3. In this new situation the probability computation requires, not four, but eight terms. For a Markov

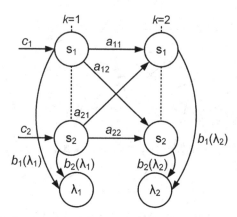

Figure 3.5 Hidden Markov model with $m = 2$ evolved up to time $k = 2$.

model with m hidden states, the number of different hidden states combinations, for k up to N, is m^N.

Assume a particular hidden states path sequence, from the universe of all the m^N possibilities. Let $Q_l = \{q_1, \cdots, q_N\}$, for $1 \leq l \leq m^N$ and $q_k \in \{s_1, \cdots, s_m\}$, be this sequence. It is worth to highlight that all the m^N sequences are different among themselves. That is, for all $i \neq l$, $Q_i \neq Q_l$. The probability of observing Λ_N, given a particular hidden states sequence Q_l, can be computed as:

$$P\left(\Lambda_N | Q_l\right) = \prod_{k=1}^{N} P\left(\lambda_k | q_k\right) = \prod_{k=1}^{N} b_{\delta_k}(\lambda_k) \qquad (3.19)$$

where δ_k is the state index pointed out by q_k. For example if $q_k = s_5$, then $\delta_k = 5$. On the other hand, the probability of observing the exact hidden sequence Q_l is:

$$P\left(Q_l\right) = P\left(q_1\right) P\left(q_2 | q_1\right) \cdots P\left(q_N | q_{N-1}\right)$$
$$= c_{\delta_1} \cdot \prod_{k=2}^{N} a_{\delta_{k-1}\delta_k} \qquad (3.20)$$

where a_{ij} is the element at i^{th} row and j^{th} column of $\mathbf{A}$. From expressions (3.18) and (3.20) the probability of obtaining $P\left(\Lambda_N | Q_l\right)$ and $P\left(Q_l\right)$ is defined as:

$$P\left(\Lambda_N | Q_l \wedge Q_l\right) = c_{\delta_1} b_{\delta_1}\left(\lambda_1\right) \cdot \prod_{k=2}^{N} a_{\delta_{k-1}\delta_k} b_{\delta_k}\left(\lambda_k\right) \qquad (3.21)$$
$$= P\left(\Lambda_N | Q_l\right) \cdot P\left(Q_l\right)$$

As it can be seen, a total of $2N - 1$ products are required to produce the above probability. In the end, the probability to observe Λ_N is given by:

$$P\left(\Lambda_N\right) = \sum_{l=1}^{m^n} P\left(\Lambda_N | Q_l \wedge Q_l\right) \qquad (3.22)$$

Besides the m^N sums, the computation of $P\left(\Lambda_N\right)$, requires a total of $m^N(2N-1)$ products. Notice that, even for modest size models and under small observed sequences, the amount of operations required to compute $P(\Lambda_N)$ leads to a severe computational load. As an example, assuming a three hidden states model and five observations, it is necessary to carry out 2277 products and 253 sums in order to compute the observation sequence probability.

The good news is that this computational load can be dramatically decreased. This reduction is obtained by recycling calculations previously carried out and just update the probability when a new observation is added. The strategy conceived for this task is designated by "Forward Algorithm".

The mechanics behind the efficient operation of the forward algorithm can be easily explained. In order to do that, take a look at Figure 3.6 where the time evolution of a Markov chain with m hidden states is represented.

The probability of obtaining λ_1 as the first observation is calculated as follows:

$$P\left(\lambda_1\right) = P\left(\lambda_1 | q_1 = s_1\right) P\left(q_1 = s_1\right) + \cdots + P\left(\lambda_1 | q_1 = s_m\right) P\left(q_1 = s_m\right)$$
$$= \sum_{i=1}^{m} P\left(\lambda_1 | q_1 = s_i\right) P\left(q_1 = s_i\right) = \sum_{i=1}^{m} P\left(\lambda_1 \wedge q_1 = s_i\right)$$

Given the definition of the matrices $\mathbf{A}$, $\mathbf{B}$ and $\mathbf{I}$, the previous expression can be rewritten as:

$$P\left(\lambda_1\right) = \sum_{i=1}^{m} b_i\left(\lambda_1\right) c_i$$

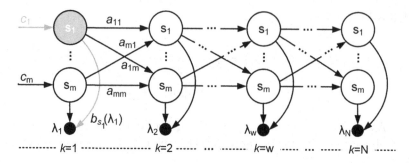

Figure 3.6 Evolution, as a function of the time k, of a Markov model with m hidden states.

Now, denoting the product $b_i(\lambda_1) c_i$ by $\alpha_1(i)$, the following equivalent expression is obtained:

$$P(\lambda_1) = \sum_{i=1}^{m} \alpha_1(i)$$

Moreover, the probability of obtaining the sequence $\Lambda_2 = \{\lambda_1 \lambda_2\}$ can be determined by:

$$P(\lambda_2 \wedge \lambda_1) = P(\lambda_2 | \lambda_1) \cdot P(\lambda_1) \tag{3.23}$$

where,

$$
\begin{aligned}
P(\lambda_2 | \lambda_1) &= P(\lambda_2 | q_2 = s_1) P(q_2 = s_1 | \lambda_1) + \cdots \\
&\quad + P(\lambda_2 | q_2 = s_m) P(q_2 = s_m | \lambda_1) \\
&= \sum_{i=1}^{m} P(\lambda_2 | q_2 = s_i) P(q_2 = s_i | \lambda_1)
\end{aligned}
\tag{3.24}
$$

The quantity $P(q_2 = s_i | \lambda_1)$ refers to the probability that, at the present, the current hidden state is s_i and λ_1 has been observed as the visible state during the previous instant. This probability can be decomposed as follows:

$$
\begin{aligned}
P(q_2 = s_i | \lambda_1) &= P(q_2 = s_i | q_1 = s_1) P(q_1 = s_1 | \lambda_1) + \cdots \\
&\quad + P(q_2 = s_i | q_1 = s_m) P(q_1 = s_m | \lambda_1) \\
&= \sum_{j=1}^{m} P(q_2 = s_i | q_1 = s_j) P(q_1 = s_j | \lambda_1)
\end{aligned}
\tag{3.25}
$$

On the other hand, $P(q_1 = s_j | \lambda_1)$ represents the probability of the hidden state, at time instant $k = 1$, being s_j given that the state λ_1 was observed. This probability can alternatively be represented, resorting to Bayes theorem, as:

$$P(q_1 = s_j | \lambda_1) = \frac{P(\lambda_1 | q_1 = s_j) P(q_1 = s_j)}{P(\lambda_1)} \tag{3.26}$$

Replacing this equality into expression (3.25) yields:

$$
\begin{aligned}
P(q_2 = s_i | \lambda_1) &= \sum_{j=1}^{m} P(q_2 = s_i | q_1 = s_j) \frac{P(\lambda_1 | q_1 = s_j) P(q_1 = s_j)}{P(\lambda_1)} \\
&= \sum_{j=1}^{m} a_{ji} \frac{b_j(\lambda_1) \cdot c_j}{P(\lambda_1)}
\end{aligned}
\tag{3.27}
$$

Since $b_j(\lambda_1) \cdot c_j$ is nothing more than $\alpha_1(j)$, the previous expression becomes,

$$P(q_2 = s_i | \lambda_1) = \sum_{j=1}^{m} a_{ji} \frac{\alpha_1(j)}{P(\lambda_1)} \tag{3.28}$$

Replacing this last expression into (3.24) leads to:

$$P(\lambda_2 | \lambda_1) = \sum_{i=1}^{m} b_i(\lambda_2) \sum_{j=1}^{m} a_{ji} \frac{\alpha_1(j)}{P(\lambda_1)} \tag{3.29}$$

which makes $P(\lambda_2 \wedge \lambda_1)$ being equal to:

$$P(\lambda_2 \wedge \lambda_1) = \sum_{i=1}^{m} b_i(\lambda_1) \sum_{j=1}^{m} a_{ji} \alpha_1(j) \tag{3.30}$$

By carrying out the assignment,

$$\alpha_2(i) = b_i(\lambda_2) \sum_{j=1}^{m} a_{ji} \alpha_1(j) \tag{3.31}$$

one obtains:

$$P(\Lambda_2) = \sum_{i=1}^{m} \alpha_2(i) \tag{3.32}$$

It is fundamental to notice that $\alpha_2(i)$ is equal to $P(\Lambda_2 \wedge q_2 = s_i)$. Thus, taking into consideration (2.33), expression (3.31) can be written as:

$$\alpha_2(i) = P(\lambda_2|q_2 = s_i)P(q_2 = s_i|\lambda_1)P(\lambda_1) \tag{3.33}$$

Due to the independence between λ_2 and λ_1, the previous equality is equivalent to:

$$\alpha_2(i) = P(\lambda_2|q_2 = s_i \wedge \lambda_1)P(q_2 = s_i|\lambda_1)P(\lambda_1) \tag{3.34}$$

And, by applying the chain rule presented in (2.30), leads to:

$$\alpha_2(i) = P(\lambda_2 \wedge \lambda_1 \wedge q_2 = s_i) \tag{3.35}$$

that is,

$$\alpha_2(i) = P(\Lambda_2 \wedge q_2 = s_i) \tag{3.36}$$

Extrapolating this way of reasoning to a new observation λ_3, $P(\Lambda_3)$ can now be determined by:

$$P(\lambda_3 \wedge \lambda_2 \wedge \lambda_1) = P(\lambda_3|\lambda_2\lambda_1)P(\Lambda_2) \tag{3.37}$$

where,

$$\begin{aligned} P(\lambda_3|\lambda_1\lambda_2) &= P(\lambda_3|q_3 = s_1) \cdot P(q_3 = s_1|\lambda_1\lambda_2) + \cdots \\ &+ P(\lambda_3|q_3 = s_m) \cdot P(q_3 = s_m|\lambda_1\lambda_2) \\ &= \sum_{i=1}^{m} P(\lambda_3|q_3 = s_i) \cdot P(q_3 = s_i|\lambda_1\lambda_2) \end{aligned} \tag{3.38}$$

On the other hand,

$$\begin{aligned} P(q_3 = s_i|\lambda_1\lambda_2) &= P(q_3 = s_i|q_2 = s_1)P(q_2 = s_1|\lambda_1\lambda_2) + \cdots \\ &+ P(q_3 = s_i|q_2 = s_m)P(q_2 = s_m|\lambda_1\lambda_2) \\ &= \sum_{j=1}^{m} P(q_3 = s_i|q_2 = s_j) \cdot P(q_2 = s_j|\lambda_1\lambda_2) \end{aligned} \tag{3.39}$$

Applying Bayes theorem to $P(q_2 = s_j|\lambda_1\lambda_2)$ one obtains:

$$\begin{aligned} P(q_2 = s_j|\lambda_1\lambda_2) &= \frac{P(\lambda_1\lambda_2 \wedge q_2 = s_j)}{P(\Lambda_2)} \\ &= \frac{\alpha_2(j)}{P(\Lambda_2)} \end{aligned} \tag{3.40}$$

and replacing (3.40) in (3.39), one gets:

$$P\left(q_3 = s_i \middle| \lambda_1 \lambda_2\right) = \sum_{j=1}^{m} a_{ji} \frac{\alpha_2(j)}{P(\Lambda_2)} \tag{3.41}$$

which, by using equality (3.38), leads to:

$$P\left(\lambda_3 \middle| \lambda_1 \lambda_2\right) = \sum_{i=1}^{m} b_i(\lambda_3) \sum_{j=1}^{m} a_{ji} \frac{\alpha_2(j)}{P(\Lambda_2)} \tag{3.42}$$

resuming to,

$$P(\Lambda_3) = \sum_{i=1}^{m} b_i(\lambda_3) \sum_{j=1}^{m} a_{ji} \alpha_2(j) \tag{3.43}$$

By defining:

$$\alpha_3(i) = b_i(\lambda_3) \left(\sum_{j=1}^{m} a_{ji} \alpha_2(j) \right)$$

then,

$$P(\Lambda_3) = \sum_{i=1}^{m} \alpha_3(i) \tag{3.44}$$

Finally, taking into account the pattern that has been revealed, it is possible to establish, generically and for an arbitrary number of observations N, the following expression:

$$P(\Lambda_N) = P\left(\lambda_N \middle| \lambda_1 \cdots \lambda_{N-1}\right) = \sum_{i=1}^{m} \alpha_N(i) \tag{3.45}$$

where,

$$\alpha_N(i) = b_i(\lambda_N) \left(\sum_{j=1}^{m} a_{ji} \alpha_{N-1}(j) \right) \tag{3.46}$$

Particular attention should be given to:

$$\alpha_1(i) = b_i(\lambda_1) c_i \tag{3.47}$$

These last three equations constitute the backbone of the forward algorithm. The application of this algorithm, to compute the probability of an output sequence, involves, after an initiation stage governed by (3.47), the recursive execution of,

$$\alpha_k (i) = b_i (\lambda_k) \left(\sum_{j=1}^{m} a_{ji} \alpha_{k-1} (j) \right) \qquad (3.48)$$

for $1 \leq k \leq N$ and $1 \leq i \leq m$.

To conclude, it is worth to notice that the forward probability $\alpha_k(i)$ represents the probability of observing the sequence Λ_k and, simultaneously, the active hidden state at k be s_i. Algebraically this statement is represented as:

$$\begin{aligned} \alpha_k(i) &= P(\lambda_1, \cdots, \lambda_k \wedge q_k = s_i) \\ &= P(\Lambda_k \wedge q_k = s_i) \end{aligned} \qquad (3.49)$$

All the algorithm phases are summarized by the following three steps:

Initialization: For $1 \leq i \leq m$ do,

$$\alpha_1 (i) = b_i (\lambda_1) c_i \qquad (3.50)$$

Recursion: For $1 \leq k \leq N$ and $1 \leq i \leq m$ do,

$$\alpha_N (i) = b_i (\lambda_N) \left(\sum_{j=1}^{m} a_{ji} \alpha_{N-1} (j) \right) \qquad (3.51)$$

Conclusion: In the end, the probability of observing Λ_N is obtained by:

$$P (\Lambda_N) = P (\lambda_N | \lambda_1 \cdots \lambda_{N-1}) = \sum_{i=1}^{m} \alpha_N (i) \qquad (3.52)$$

This algorithm, coded as a MATLAB® function, is presented at Listing 3.1.

Listing 3.1 Forward Algorithm coded for MATLAB®.

```
1 function varargout=forward_algorithm(A,B,O,I)
2 % Forward Algorithm for discrete hidden Markov Models with 'm' hidden
3 % states, 'n' observable states and 'N' observations.
4 %    A - mxm (state transition matrix)
5 %    B - mxn (confusion matrix)
6 %    O - 1xN (observations vector)
7 %    c - 1xm (initial probabilities vector)
```

```
 8 %
 9 % Usage:
10 %    [Alfa]=forward_algorithm(A,B,O,I)
11 %    [Alfa,LogLik]=forward_algorithm(A,B,O,I)
12 %
13 % Where:
14 %    Alfa - Partial probability matrix P(Lambda_k ^ q_k).
15 %    LogLik - Log-likelihood of O.
16
17 [m,n]=size(B);
18 N=length(O);
19
20 %% Initialization
21 Alfa=zeros(N,m);
22 for k=1:m
23     Alfa(1,k)=I(k)*B(k,O(1));
24 end
25
26 %% Recursion
27 for l=2:N,
28     for k=1:m,
29         S=0;
30         for i=1:m,
31             S=S+A(i,k)*Alfa(l-1,i);
32         end
33         Alfa(l,k)=B(k,O(l))*S;
34     end
35 end
36
37 % Probability of observing O
38 P=sum(Alfa(N,:));
39
40 % function return
41 if nargout==1,
42     varargout={Alfa};
43 else
44     varargout(1)={Alfa};varargout(2)={log(P)};
45 end
```

The following section presents a numeric method closely related to the "forward algorithm". This alternative technique, known as "backward algorithm", can be used to efficiently compute the probability of obtaining a fraction of the observed sequence.

3.2.2 THE BACKWARD ALGORITHM

The "backward algorithm" is a numerically efficient method used to obtain the probability of getting, at an arbitrary time step k, a fraction of the observed sequence. That is, the backward algorithm involves the computation of $P(\Lambda_{k+1 \to N} | q_k = s_i)$ where $\Lambda_{k+1 \to N}$ denotes the set of observations taken from time $k+1$ up to N. That is, $\{\lambda_{k+1} \cdots \lambda_N\}$. This conditional probability will be denoted, in a more compact form, as $\beta_k(i)$.

As for the forward algorithm, there is a way to calculate this probability recursively. In order to gain insight on how this can be done, let's consider a particular structure consisting of two hidden states, s_1 and s_2, as represented in Figure 3.7. Starting from $P(\lambda_4 | q_3)$, it is possible to observe that this probability consists on the sum of the probability to reach the observation λ_4 through two distinct paths: one that goes through s_1 and another through s_2. This situation can be written as follows:

$$
\begin{aligned}
\beta_3(1) =& P(\lambda_4 | q_3 = s_1) \\
=& P(\lambda_4 | q_4 = s_1) P(q_4 = s_1 | q_3 = s_1) + \\
& P(\lambda_4 | q_4 = s_2) P(q_4 = s_2 | q_3 = s_1) \\
=& b_1(\lambda_4) a_{11} + b_2(\lambda_4) a_{12}
\end{aligned}
\tag{3.53a}
$$

$$
\begin{aligned}
\beta_3(2) =& P(\lambda_4 | q_3 = s_2) \\
=& P(\lambda_4 | q_4 = s_1) P(q_4 = s_1 | q_3 = s_2) + \\
& P(\lambda_4 | q_4 = s_2) P(q_4 = s_2 | q_3 = s_2) \\
=& b_1(\lambda_4) a_{21} + b_2(\lambda_4) a_{22}
\end{aligned}
\tag{3.53b}
$$

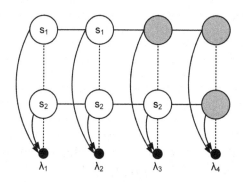

Figure 3.7 Representation of the reverse probability, β, in $k = 3$, for a model with two hidden states.

Notice that, for the generic case of a model with m hidden states and N observations, the previous expressions can be represented as:

$$
\beta_{N-1}(i) = \sum_{j=1}^{m} b_j(\lambda_N) a_{ij}, \ i = 1, \ldots, m
\tag{3.54}
$$

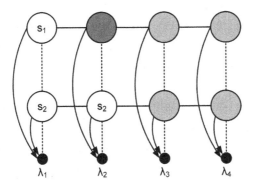

Figure 3.8 Representation of the reverse probability $\beta_2(i)$.

The probability $P(\lambda_3\lambda_4 | q_2 = s_i)$ of observing λ_3 in $k = 3$ and λ_4 in $k = 4$, given that, in $k = 2$, and the active hidden state is s_i for $i = \{1, 2\}$, is illustrated in Figure 3.8.

The backward probability for $k = 3$ and $s_i = 1$ is given by:

$$\beta_2(1) = P(\lambda_3\lambda_4 | q_2 = s_1) \tag{3.55}$$

By taking into consideration the property defined at (2.34), the previous expression is expanded as:

$$
\begin{aligned}
\beta_2(1) &= P(\lambda_3|s_1) \cdot P(\lambda_4|s_1) \\
&= P(\lambda_3|s_1) \cdot \Big[b_1(\lambda_4)\, a_{11}a_{11} + b_1(\lambda_4)\, a_{21}a_{12} + \\
&\qquad\qquad + b_2(\lambda_4)\, a_{12}a_{11} + b_2(\lambda_4)\, a_{22}a_{12} \Big] \\
&= b_1(\lambda_3) \cdot \Big[a_{11}(b_1(\lambda_4)\, a_{11} + b_2(\lambda_4)\, a_{12}) + \\
&\qquad\qquad + a_{12}(b_1(\lambda_4)\, a_{21} + b_2(\lambda_4)\, a_{22}) \Big]
\end{aligned}
\tag{3.56}
$$

The same idea can be applied to $\beta_2(2)$ which leads to the following equalities:

$$\beta_2(1) = b_1(\lambda_3) \cdot \left[a_{11}\beta_3(1) + a_{12}\beta_3(2) \right]$$

$$= \sum_{j=1}^{2} b_1(\lambda_3) a_{1j}\beta_3(j)$$

$$\beta_2(2) = b_1(\lambda_3) \cdot \left[a_{21}\beta_3(1) + a_{22}\beta_3(2) \right]$$

$$= \sum_{j=1}^{2} b_1(\lambda_3) a_{2j}\beta_3(j)$$

(3.57)

Again, in the case of m hidden states and N observations, the above expressions condense into:

$$\beta_{N-2}(i) = \sum_{j=1}^{m} b_j(\lambda_{N-1}) a_{ij}\beta_{N-1}(j), \ i = 1, \ldots, m \tag{3.58}$$

Taking into consideration the previous two cases, it is possible to detect a pattern which enable us to define a general way to recursively calculate the backward probability for any time instant k. In particular, the backward probability can be computed from:

$$\beta_{k-1}(i) = \sum_{j=1}^{m} b_j(\lambda_k) a_{ij}\beta_k(j) \tag{3.59}$$

where, $i = 1, \ldots, m$ and $k = 1, \ldots, N$.

To complete the above expression, it is necessary to define $\beta_N(j)$. Comparing the previous equation, with the one in (3.54), for $k = N$ it is possible to verify that:

$$\beta_N(i) = 1, \text{for } i = 1, \cdots, m \tag{3.60}$$

In the previous section, it was seen that $\alpha_k(i)$ refers to the joint probability of observing the set Λ_k and, at the same time, the active hidden state at k to be s_i. That is,

$$\alpha_k(i) = P(\lambda_1 \cdots \lambda_k \wedge q_k = s_i)$$
$$= P(\Lambda_k \wedge q_k = s_i) \tag{3.61}$$

Additionally, the probability of observing Λ_k could be determined by the sum:

$$P(\Lambda_k) = \sum_{i=1}^{m} \alpha_k(i) \tag{3.62}$$

48

On the other hand, in this section, the notion of conditional probability $\beta_k(i)$ has been developed as being the probability of partial observation $\lambda_{k+1} \cdots \lambda_N$ given that the active hidden state, at time step k, is s_i. That is,

$$\beta_k(i) = P\left(\lambda_{k+1} \cdots \lambda_N \middle| q_k = s_i\right)$$
$$= P\left(\Lambda_{k+1 \to N} \middle| q_k = s_i\right)$$

(3.63)

where $\Lambda_{k+1 \to N}$ refers to observations taken from the time instants in the interval extending from $k + 1$ to N. It is also straightforward to see that:

$$P\left(\Lambda_{k+1 \to N}\right) = \sum_{i=1}^{m} \beta_k(i)$$

(3.64)

To demonstrate the above equality, it is enough to see that, due to the independence between observations, the probability $P\left(\lambda_{k+1} \cdots \lambda_N\right)$ can be written as:

$$P\left(\lambda_{k+1} \cdots \lambda_N\right) = P\left(\lambda_{k+1} \cdots \lambda_{N-1}\right) \cdot P\left(\lambda_N\right)$$
$$= P\left(\lambda_{k+1} \cdots \lambda_{N-1}\right) \sum_{i=1}^{m} P\left(\lambda_N \middle| q_N = s_i\right)$$

(3.65)

Just by reordering the previous expression yields:

$$P\left(\lambda_{k+1} \cdots \lambda_N\right) = \sum_{i=1}^{m} P\left(\lambda_{k+1} \cdots \lambda_{N-1}\right) P\left(\lambda_N \middle| q_N = s_i\right)$$

(3.66)

which, due to the independence between $\Lambda_{k+1 \to N-1}$ and q_N, can be further written as:

$$P\left(\lambda_{k+1} \cdots \lambda_N\right) = \sum_{i=1}^{m} P\left(\lambda_{k+1} \cdots \lambda_N \middle| q_k = s_i\right)$$
$$= \sum_{i=1}^{m} \beta_k(i)$$

(3.67)

Similarly to the forward algorithm, the backward algorithm can be summarized by the following three iterative steps:

Initialization: For $i = 1, \cdots, m$ do,

$$\beta_N(i) = 1$$

(3.68)

49

Recursion: For $i = 1, \ldots, m$ and $k = 1, \ldots, N$ do,

$$\beta_{k-1}(i) = \sum_{j=1}^{m} b_j(\lambda_k) a_{ij} \beta_k(j) \tag{3.69}$$

Conclusion: At the end, the probability $P(\Lambda_{k+1 \to N})$ is computed by:

$$P(\lambda_{k+1} \cdots \lambda_N) = \sum_{i=1}^{m} \beta_k(i) \tag{3.70}$$

Listing 3.2 presents a MATLAB® function that implements the above described three steps of the backward algorithm. The state transition matrix $\mathbf{A}$ and the emissions matrix $\mathbf{B}$ are sent as arguments and the code returns a matrix representing $\beta_k(i)$ for $i = 1, \cdots, m$ and $k = 1, \cdots, N$.

Listing 3.2 Backward algorithm coded for MATLAB®.

```
1  function Beta=backward_algorithm(A,B,O)
2
3  % Backward Algorithm for discrete hidden Markov Models with 'm' hidden
4  % states, 'n' observable states and 'N' observations.
5  %    A - mxm (state transition matrix)
6  %    B - mxn (confusion matrix)
7  %    O - 1xN (observations vector)
8
9  [m,n]=size(B);
10 N=length(O);
11
12 %% Initialization
13 Beta=zeros(N,m);
14 for k=1:m
15     Beta(N,k)=1;
16 end
17
18 %% Recursion
19 for t=N-1:-1:1,
20     for i=1:m,
21         Beta(t,i)=0;
22         for j=1:m,
23             Beta(t,i)=Beta(t,i)+A(i,j)*B(j,O(t+1))*Beta(t+1,j);
24         end
25     end
26 end
```

Just before concluding this section, an interesting fact is highlighted: the backward and forward probabilities can be used together to compute $P(\Lambda_N)$. As a matter of fact, the probability of observing Λ_N can be obtained from:

$$P(\Lambda_N) = \sum_{i=1}^{m} \alpha_k(i) \beta_k(i) \tag{3.71}$$

Notice that this relationship holds for any arbitrary k between 1 and m.

The next section introduces the Viterbi algorithm. This algorithm can be applied to many practical situations outside the hidden Markov models such as in IEEE 802.11 wireless LAN's [19] and many other telecommunication protocols. Within the hidden Markov model paradigm, the Viterbi algorithm allows to compute the likeliest hidden states sequence from a set of observations. Details regarding this method will be addressed in the following section.

3.2.3 THE VITERBI ALGORITHM

Usually, when dealing with hidden Markov models, three fundamental algorithms are considered the forward, backward and Viterbi algorithms. The former two were presented within the previous subsections and the latter will be addressed along the present one. The aim of the Viterbi algorithm is to provide a computationally efficient method to obtain the likeliest hidden states sequence from a set of N observations $\{\lambda_1, \cdots, \lambda_N\}$.

Let's assume that it is required to determine which one of the two states, s_1 or s_2, is more likely to find actually active considering that the state λ_1 is observed. The answer seems simple: just choose the state whose path to λ_1 is the most probable. Figure 3.9 illustrates this idea.

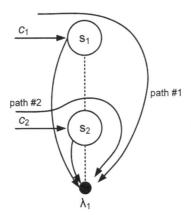

Figure 3.9 The two possible paths in the initial situation of a Markov model with two hidden states.

If path "1" is defined as $P(\lambda_1|c_1)$ and path "2" as $P(\lambda_1|c_2)$ then, in the absence of more information, the most likely active hidden state is the one found on the path whose probability leading to λ_1 is higher. For example, if $c_1 = 0.5$, $c_2 = 0.5$, $P(\lambda_1|q_1 = s_1) = 0.2$ and $P(\lambda_1|q_1 = s_2) = 0.3$ then $P(\lambda_1|c_1) = 0.1$ and $P(\lambda_1|c_2) = 0.15$. In this case, path "2" is more likely to lead to λ_1 than path "1". This may imply that, in theory, it is the state s_2 that is active at the instant the observation λ_1 was registered. Notice that the startup situation of the model is singular since only one path passes through each hidden state.

Consider now, the addition of a second observation, λ_2. Which of the two possible states, s_1 or s_2, is now active? Figure 3.10 represents this new situation with transition probability values arbitrarily chosen.

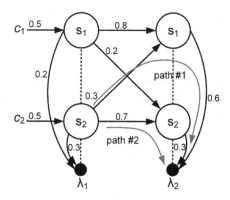

Figure 3.10 The two possible paths to obtain λ_2 given that the hidden state in $k = 1$ is s_2.

If it is assumed that the active state, associated to the first observation, is s_2 (since it is the one that maximizes the probability) then the two possible paths for the observation λ_2 are:

$$P(\lambda_2|q_2 = s_1) P(q_2 = s_1|q_1 = s_2) = 0.18$$
$$P(\lambda_2|q_2 = s_2) P(q_2 = s_2|q_1 = s_2) = 0.21 \tag{3.72}$$

Then, the active state with the highest probability at the instant $k = 2$, is the state s_2. It is simple to confirm that the probability of the sequence $\lambda_1\lambda_2$ to be generated by the hidden state sequence s_2s_2 is equal to the product $0.21 \times 0.15 = 0.0315$. That is,

$$P\left(\lambda_1\lambda_2|s_2s_2\right) =$$
$$P\left(\lambda_1|q_1 = s_2\right)P\left(q_1 = s_2\right)P\left(\lambda_2|q_2 = s_2\right)P\left(q_2 = s_2|q_1 = s_2\right) \quad (3.73)$$
$$= 0.0315$$

What will be the probability value if a different path has been chosen instead?

To answer this question, Table 3.1 presents the probabilities associated with the observations λ_1 and λ_2 considering all the four possible paths.

Table 3.1

Probability of $\lambda_1\lambda_2$ according to the four possible paths for the model represented in the Figure 3.10.

Hidden Sequence	$P\left(\lambda_1\lambda_2|\text{sequence}\right)$
s_1s_1	$0.5 \times 0.2 \times 0.8 \times 0.6 = 0.048$
s_1s_2	$0.5 \times 0.2 \times 0.2 \times 0.4 = 0.008$
s_2s_1	$0.5 \times 0.3 \times 0.3 \times 0.6 = 0.027$
s_2s_2	$0.5 \times 0.3 \times 0.7 \times 0.3 = 0.0315$

From this table, it is possible to conclude that there is a larger probability that the observed sequence has been generated by the hidden sequence s_1s_1 than by the sequence s_2s_2. This result seems to contradict the previous reasoning. The problem was that a deterministic decision has been made on probabilistic information. By assuming the first state as s_2, half of the likelihood paths leading to λ_2 would be eliminated. Hence, the decision regarding the most probable sequence of the hidden states, should be done adaptively. This means that, as new observations are available, more information is gathered regarding the dynamic behavior of the model. However, there is a problem with this approach: the number of paths to be established, each time an observation is added, increases exponentially. More specifically, for a set of N observations, and considering an ergodic network with m hidden states, the number of paths to be considered increases according to m^N.

As with the forward and backward algorithms, there is a way to circumvent this computational complexity scaling issue by including past calculations in order to speed up the process. That is, it uses the most probable paths for each of the hidden states up to the immediately previous time instant. This process is taken iteratively and is known as the Viterbi algorithm [34].

In order to understand how this algorithm operates imagine a new observation λ_3 associated with the same hidden Markov model. Figure 3.11 illustrates this situation.

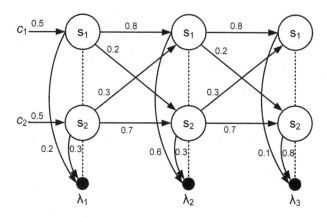

Figure 3.11 Adding a new observation, λ_3, to the model shown in the Figure 3.10.

In this case, the calculation of the likeliest hidden states sequence involves the analysis of 8 different paths. The set of those sequences, as well as their respective probabilities, is found in Table 3.2.

Table 3.2

Probability of $\lambda_1\lambda_2\lambda_3$ according to the eight possible paths.

Hidden Sequence	$P\left(\lambda_1\lambda_2\lambda_3 \vert \text{sequence}\right)$
$s_1 s_1 s_1$	$0.5 \times 0.2 \times 0.8 \times 0.6 \times 0.8 \times 0.1 = 0.00384$
$s_1 s_1 s_2$	$0.5 \times 0.2 \times 0.8 \times 0.6 \times 0.2 \times 0.8 = 0.00768$
$s_1 s_2 s_1$	$0.5 \times 0.2 \times 0.2 \times 0.3 \times 0.3 \times 0.1 = 0.00018$
$s_1 s_2 s_2$	$0.5 \times 0.2 \times 0.2 \times 0.3 \times 0.7 \times 0.8 = 0.00336$
$s_2 s_1 s_1$	$0.5 \times 0.3 \times 0.3 \times 0.6 \times 0.8 \times 0.1 = 0.00216$
$s_2 s_1 s_2$	$0.5 \times 0.3 \times 0.3 \times 0.6 \times 0.2 \times 0.8 = 0.00432$
$s_2 s_2 s_1$	$0.5 \times 0.3 \times 0.7 \times 0.3 \times 0.3 \times 0.1 = 0.00095$
$s_2 s_2 s_2$	$0.5 \times 0.3 \times 0.7 \times 0.3 \times 0.7 \times 0.8 = 0.01764$

Now, the number of patterns to analyze has doubled. Nevertheless, consider the table rows, taken in groups of two, and compare them with the ones of Table 3.1. This procedure is illustrated in Figure 3.12.

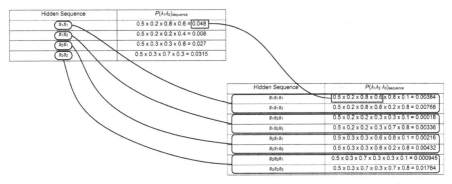

Figure 3.12 Association of the values of the probabilities relative to all possible paths obtained from the models presented in the Figures 3.9 and 3.10.

It can be seen that, for each pair of adjacent rows, the probability value of the sequence obtained up to $k = 2$ does not change. Thus, in the presence of a new observation, it is not necessary to perform all the products again. The values previously determined for $k = 2$ can be used for $k = 3$, simply multiplying them by the transition probability and by the observation probability. Although the number of operations, per path, is reduced with this strategy, the problem of exponential number of paths, which arises with each new observation, continues to exist. The solution to this problem is very simple and elegant. First of all it should be remarked that there are four groups of two rows in the Table 3.2 which, as it is known, represent the four possible paths for attaining the observed sequence. Note also that two of the paths go through s_1 at $k = 2$ and other two through s_2 at the same time instant. When a new observation is gathered, the probabilistic arrangement is multiplied by each of the paths associated with each hidden state. Now, if one of the paths already has a larger associated probability, it is logical that this advantage is forwarded for the new probability calculation. Thus, from the calculation at $k = 2$, it is enough to retain one of the two main paths associated with each node. In this case, the information associated with the path $s_1 s_1$ and the path $s_2 s_2$. This strategy is illustrated in the following sequence of Figures 3.13 to 3.14.

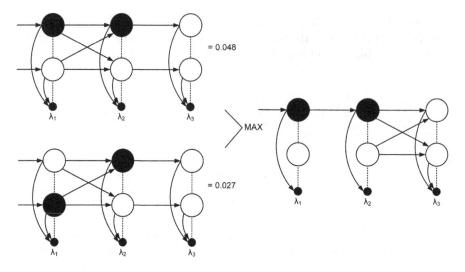

Figure 3.13 At each instant k is enough to retain one of the two main paths associated with each node: analysis for s_1 in $k = 2$.

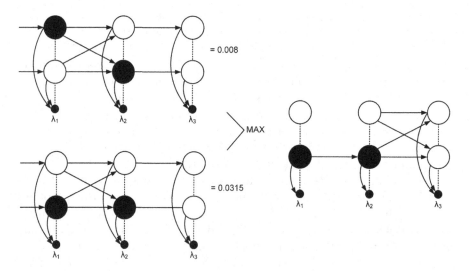

Figure 3.14 At each instant k is enough to retain one of the two main paths associated with each node: analysis for s_2 in $k = 2$.

Taking into consideration only the best paths associated with each node, when a new observation arrives, only the evaluation of m^2 paths is required. For example, in the current example, at $k = 2$ the results provided in Table 3.3 are obtained.

Table 3.3

Probability of observing $\lambda_1 \lambda_2$ according to the four possible paths.

Hidden Sequence	$P(\lambda_1 \lambda_2 \| \text{sequence})$
$s_1 s_1$	$0.5 \times 0.2 \times 0.8 \times 0.6 = 0.048$
$s_2 s_2$	$0.5 \times 0.3 \times 0.7 \times 0.3 = 0.0315$

Of course, this strategy can be extended to $k = 3$. In this situation it is only necessary to evaluate four situations. Table 3.4 summarizes the values obtained for each of the possible four paths.

Table 3.4

Probability of $\lambda_1 \lambda_2 \lambda_3$ according to the four possible paths.

Hidden Sequence	$P(\lambda_1 \lambda_2 \lambda_3 \| \text{sequence})$
$s_1 s_1 s_1$	$0.5 \times 0.2 \times 0.8 \times 0.6 \times 0.8 \times 0.1 = 0.00384$
$s_1 s_1 s_2$	$0.5 \times 0.2 \times 0.8 \times 0.6 \times 0.2 \times 0.8 = 0.00768$
$s_2 s_2 s_1$	$0.5 \times 0.3 \times 0.7 \times 0.3 \times 0.3 \times 0.1 = 0.00095$
$s_2 s_2 s_2$	$0.5 \times 0.3 \times 0.7 \times 0.3 \times 0.7 \times 0.8 = 0.01764$

Following this discussion, it is easy to predict that, for $k = 4$, it is enough to calculate the situations presented in Table 3.5. The Viterbi algorithm takes advantage of the described reasoning. However, before presenting it, let's define two new variables: $\rho_j(i)$ and $\psi_j(i)$. The former refers to the maximum probability of obtaining the observable sequence from $k = 1$ to $k = j$ supposing that, at $k = j$, the active hidden state is s_i. That is $\rho_j(i) = \max\{P(\lambda_1 \cdots \lambda_j \wedge q_j = s_i)\}$. On the other hand $\psi_j(i)$ works as a pointer and defines, during iteration j, which hidden states will lead, at $k = j - 1$, to a larger ρ_k value. That is, for an arbitrary instant k, and for $i, j = 1, \cdots m$,

$$\psi_k(i) = \arg\max_j \{\rho_{k-1}(j) \cdot a_{ji} \cdot b_i(\lambda_k)\} \tag{3.74}$$

Table 3.5

Probability of $\lambda_1\lambda_2\lambda_3\lambda_4$ according to the four possible paths.

Hidden Sequence	$P(\lambda_1\lambda_2\lambda_3\lambda_4 \vert \text{sequence})$
$s_1 s_1 s_1 s_1$	$0.00384 \times P(\lambda_4 \vert s_1) P(s_1 \vert s_1)$
$s_1 s_1 s_1 s_2$	$0.00384 \times P(\lambda_4 \vert s_2) P(s_2 \vert s_1)$
$s_2 s_2 s_2 s_1$	$0.01764 \times P(\lambda_4 \vert s_1) P(s_1 \vert s_2)$
$s_2 s_2 s_2 s_2$	$0.01764 \times P(\lambda_4 \vert s_2) P(s_2 \vert s_2)$

After the introduction of these two new variables the Viterbi algorithm can be described in the following three steps:

Initialization: When $k = 1$, $\rho_1(i)$ is computed by:

$$\rho_1(i) = c_i \cdot b_i(\lambda_1), \quad i = 1, \cdots, m \tag{3.75}$$

This value is normalized in order to prevent numeric underflow. The normalized $\rho_1(i)$, denoted by $\rho_1'(i)$, is obtained by:

$$\rho_1'(i) = \frac{\rho_1(i)}{\sum_{j=1}^{m} \rho_1(j)}, \quad i = 1, \cdots, m \tag{3.76}$$

Now, since there are no active states previously active, ψ is initialized to zero. That is,

$$\psi_1(i) = 0, \quad i = 1, \cdots, m \tag{3.77}$$

Recursion: For $1 < k \leq N$ and $i, j = 1, \cdots, m$,

$$\rho_k(i) = \max \left\{ \rho_{k-1}'(j) \cdot a_{ji} \cdot b_i(\lambda_k) \right\} \tag{3.78}$$

Once again the previous values are normalized to prevent numeric underflow. This leads to:

$$\rho_k'(i) = \frac{\rho_k(i)}{\sum_{j=1}^{m} \rho_k(j)}, \quad i = 1, \cdots, m \tag{3.79}$$

and $\psi_k(i)$ is updated according to:

$$\psi_k(i) = \arg \max_j \left\{ \rho_{k-1}'(j) \cdot a_{ji} \right\}, i, j = 1, \cdots, m \tag{3.80}$$

States Revelation: The last stage in this algorithm concerns the active states revelation along a path. The active state, in the last epoch stage, is the one that exhibits higher value for $\rho_N(i)$. That is,

$$q_N = \arg\max_i \{\rho_N(i)\} \text{ for } i = 1, \cdots, m \tag{3.81}$$

The active state with highest probability to be active is obtained from the values of ψ_N. That is, the more probable path to attain q_N, from q_{N-1}, is obtained from the node pointed out by ψ_N. Additionally, for $k = N - 2$, this information is devised from $\psi_{N-1}(i)$. It resembles as if the model structure illustrated in Figure 3.15 was traversed, from front to back, following a path pointed out by the ψ variable associated to the present active state. That is,

$$q_k = \psi_{k+1}(q_{k+1}), \ k = N - 1, \cdots, 1 \tag{3.82}$$

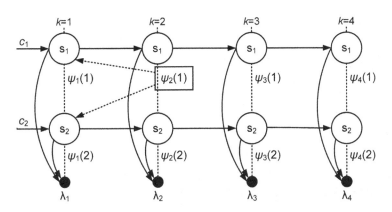

Figure 3.15 Representation on how the hidden chain is traversed, along the more probable states, by using the information contained in $\psi_k(i)$.

The MATLAB® version of the above described algorithm can be found at Listing 3.3. Once again the input arguments are, besides the observations vector, the matrices **A**, **B** and the vector **c**. This function returns the most probable state sequence in an N-dimensional **q** vector.

Listing 3.3 Viterbi algorithm coded for MATLAB®.

```
1  function q=viterbi_algorithm(A,B,O,c)
2  % Viterbi Algorithm for discrete hidden Markov Models with 'm' hidden
3  % states, 'n' observable states and 'N' observations.
4  %   A - mxm (state transition matrix)
5  %   B - mxn (confusion matrix)
6  %   O - 1xN (observations vector)
7  %   c - 1xm (initial probabilities vector)
8
9  [m,n]=size(B);
10 N=length(O);
11
12 %% Initialization
13 delta=zeros(N,m);phi=zeros(N,m);
14 t=1;
15 for k=1:m
16     delta(t,k)=c(k)*B(O(t),k);
17     phi(t,k)=0;
18 end
19
20 %% Recursion
21 for t=2:N,
22     for k=1:m,
23         for l=1:m,
24             tmp(l)=delta(t-1,l)*A(l,k)*B(k,O(t));
25         end
26         [delta(t,k),phi(t,k)]=max(tmp);
27     end
28 end
29
30 %% Path finding
31 q=zeros(N,1);
32 [~,Inx]=max(delta(N,:));
33 q(N)=Inx;
34 for k=N-1:-1:1,
35     q(k)=phi(k+1,q(k+1));
36 end
```

3.3 PROBABILITY TRANSITIONS ESTIMATION

Up to now, nothing has been said about the content of the probability matrices. That is, how each one of the elements in **A**, **B** and **c** is defined. The process of getting these values is known as the model parameters estimation and, in hidden Markov models, the parameters estimation problem boils down to the procedure of adjusting the state transition probabilities.

For a generic model with m hidden states and n observable states, the objective is to find the element values for **A** and **B** matrices and **c** vector, in such a way that the model describes, with some degree of accuracy, the dynamics of a given stochastic process. The elements of the $m \times m$ matrix **A** are the transition probabilities between the hidden states and the $m \times n$ elements of matrix **B** denote the transition probabilities

60

between the hidden and the observable states. Finally, the n-dimension *priors* vector $\mathbf{c}$, represent the initial hidden state activation probability.

The hidden Markov model will try to capture the stochastic process dynamics by using a finite training set of data generated by this indeterminate process. This training set will be used to compute a benchmark function, which will then be used to assess the hidden Markov process representation ability. Generally the "model likelihood" is used as the referred benchmark function. The model likelihood is relative to the probability of a set of gathered training data, that has been generated by a given model. Hence, the training algorithm will seek for a set of transition probabilities looking to maximize this model likelihood.

As a side note, in statistics there is a clear distinction between probability and likelihood. By definition, probability allows the prediction of unknown results based on known model parameters. On the other hand, likelihood allows the estimation of statistical model parameters based only on process observations.

Let us assume a statistical process governed by a probability density function $p(\lambda|\Theta)$ where the vector Θ concerns the process model control parameters. Suppose also a sequence of N observations,

$$\Lambda_N = \{\lambda_1, \cdots, \lambda_N\} \tag{3.83}$$

recorded from the process with the above introduced probability density function. Assuming statistical independence among observations then,

$$p(\Lambda_N|\Theta) = \prod_{i=1}^{N} p(\lambda_i|\Theta) \tag{3.84}$$

This function is commonly designated by observations likelihood and will be, from this point forward, denoted by $\mathcal{L}(\Theta|\Lambda_N)$. Notice that the likelihood assumes knowledge on the observations sequence but ignores the statistical model parameters. For this reason, the likelihood allows the estimation of the unknown model parameters Θ based only on Λ_N. This is done by finding a vector Θ that maximizes $\mathcal{L}(\Theta|\Lambda_N)$. That is, the Θ vector is obtained by solving the following optimization problem:

$$\max_{\Theta} \quad \mathcal{L}(\Theta|\Lambda_N) \tag{3.85}$$

Eventually, there may be restrictions on the parameter space Θ. For this reason, and depending on the likelihood function or the constraints, the above optimization problem can be very complex to solve. In particular, for a hidden Markov model, the likelihood computation presents several challenges. Besides being computational intensive, it describes a non-convex search space with several maxima [40]. The next section will be devoted to further analyse the profile of the likelihood function.

3.3.1   MAXIMUM LIKELIHOOD DEFINITION

Suppose that it is required to compute the parameters Θ of a model from just a set of N observations $\lambda_1, \cdots, \lambda_N$ denoted as Λ_N. One way to do this is by searching, over the parameter space Θ, and find the one that maximizes $P(\Lambda_N|\Theta)$. This strategy is referred as "maximum likelihood" since the objective is to maximize the probability that the observed sequence Λ_N has been generated by a model with parameters Θ.

Notice that, in $P(\Lambda_N|\Theta)$, the observations sequence Λ_N is known and constant. This implies that $P(\Lambda_N|\Theta)$ depends only on Θ. For this reason $P(\Lambda_N|\Theta)$ does not represent a probability distribution but only a function of Θ. In [8] this function is designated by fiducial. For this reason $P(\Lambda_N|\Theta)$ is denoted by "likelihood" in contrast with the concept of "probability" where the sequence Λ_N is the unknown and Θ is given.

It should be noted that, since there is not an universe of correct parameters for a given model, it is not possible to derive confidence intervals for the parameters. However, in the reverse problem, it is possible to establish confidence intervals for the observations. An interesting text about the different nature of probability and likelihood can be found in [15].

Now, consider a hidden Markov model with m hidden states and a particular hidden states sequence $Q_l = \{q_1, \cdots, q_k\}$ for $q_i \in \{s_1, \cdots, s_m\}$. Let Θ be the set of all probability transitions within the model. The likelihood associated to a set of observations Λ_k, given a generic hidden states sequence Q_l, is:

$$P(\Lambda_k|Q_l \wedge \Theta) = \prod_{i=1}^{k} P(\lambda_i|q_i) = \prod_{i=1}^{k} b_{\delta_i}(\lambda_i) \qquad (3.86)$$

where δ_i denotes the hidden state index pointed out by q_i. For example, if $q_i = s_3$ then $\delta_i = 3$.

The total likelihood of the observable state sequence Λ_k, is computed as a weighted average of all partial likelihoods subjected to each one of the possible m^k hidden states sequence. The weighting factors regard the occurrence probability of each of those possible paths. For example, for the generic hidden states sequence Q_l, this path probability is computed as:

$$P(Q_l) = c_{\delta_1} \prod_{i=2}^{k} a_{\delta_{i-1}\delta_i} \tag{3.87}$$

where a_{ij} is the ith row, j^{th} column element of the $\mathbf{A}$ matrix and c_i is the ith element of vector $\mathbf{c}$.

In this frame of reference, the likelihood of a generic m hidden states Markov model, for a set of k observations, is given by:

$$\mathcal{L}(\Theta|\Lambda_k) = \sum_{l=1}^{m^k} P(\Lambda_k|Q_l \wedge \Theta) \cdot P(Q_l) \tag{3.88}$$

where $Q_i = Q_j$ if, and only if, $i = j$.

The computational load, associated to the execution of the above mathematical expression, is very large and increases exponentially with both the model size and the number of observations. However, using the forward algorithm, this computational load can be greatly reduced. In particular, the model likelihood can be computed according to:

$$\mathcal{L}(\Theta|\Lambda_k) = \sum_{i=1}^{m} \alpha_k(i) \tag{3.89}$$

where $\alpha_k(i)$ regards the probability of observing the sequence Λ_k and, simultaneously, the active hidden state at time k be s_i.

Expression (3.88) is fundamental for the majority of hidden Markov model training algorithms. Any training method will try to find Θ in order to drive the model likelihood as high as possible. The celebrated Baum-Welch method, which will be reviewed in the next section, is a good example of this.

The Baum-Welch method has its roots in a family of algorithms denoted by "expectation-maximization" which have the likelihood as the objective function. However, in this book, the Baum-Welch training method will be addressed in a more informal way.

3.3.2 THE BAUM-WELCH TRAINING ALGORITHM

In a pure frequentist analysis perspective, the transition probability from hidden states s_i to s_j, denoted previously by a_{ij}, can be viewed in the following way: for a large enough number of observations, the transition probability from s_i to s_j is equal to the ratio between the number of times this transition occurs and the number of times the state s_i is visited. For example, suppose a hidden Markov model with two hidden states. Moreover, assume that in a series of consecutive observations, the state s_1 was active 200 times and, in 50 of them, the next active state was the state s_2. Hence, one can assign to a_{11} the value $\frac{150}{200} = 0.75$ and to a_{12} the value $\frac{50}{200} = 0.25$. To implement this probability estimation strategy it is therefore necessary to obtain, for each time instant k, the expected number of transitions from each hidden state to the observable state λ_k.

Let $\gamma_k(i)$ be the probability of, at time instant k, the active hidden state be s_i given the observations sequence Λ_N. That is,

$$\gamma_k(i) = P(q_k = s_i | \Lambda_N) \tag{3.90}$$

for $i = 1, \cdots, m$.

Additionally, let ν_i be the expected number of visits to state s_i during the N observations:

$$\nu_i = \sum_{k=1}^{N} \gamma_k(i) \tag{3.91}$$

Listing 3.4 presents a MATLAB® function that will be used ahead to compute the value for ν for each hidden state.

Listing 3.4 MATLAB® function for computing ν.

```
1  function nu=compute_nu(Gama,B)
2  % Return the number of visits to state i
3  % m hidden states, n output states and N observations
4  %
5  % B - m,n (confusion matrix)
6  % nu - mxn matrix
7
8  ["`,n]=size(B);
9  nu=(sum(Gama)).'*ones(1,n);    % Sum along the columns of Gamma
```

Moreover, given an arbitrary sequence of observations, the probability of having s_i as the active hidden state at time k and, at $k+1$, the active state to be s_j, for $j = 1, \cdots, m$, will be referred to as $\xi_k(i,j)$. That is,

$$\xi_k(i,j) = P(q_k = s_i \wedge q_{k+1} = s_j | \Lambda_N) \tag{3.92}$$

The estimated transition frequency, between s_i and s_j along the N observations, will be referred to as τ_{ij} and computed by:

$$\tau_{ij} = \sum_{k=1}^{N-1} \xi_k(i,j) \qquad (3.93)$$

The MATLAB® code that will be used for computing τ_{ij} is presented in Listing 3.5.

Listing 3.5 MATLAB® function for computing τ_{ij}.

```
1  function tau=compute_tau(Alfa,Beta,A,B,O)
2  % Compute tau...
3  % m hidden states, n output states and N observations
4  % Alfa - Nxm (from the forward algorithm)
5  % Beta - Nxm (from the backward algorithm)
6  % A - mxm (state transitions matrix)
7  % B - nxm (confusion matrix)
8  % O - 1xN (observations vector)
9
10 [m,~]=size(B);
11 N=length(O);
12 tau=zeros(m,m);
13 for k=1:N-1,
14     num=A.*(Alfa(k,:).'*Beta(k+1,:)).*(B(:,O(k+1))*ones(1,m)).';
15     den=ones(1,m)*num*ones(m,1);
16     tau=tau+num/den;
17 end
```

Let's also define τ_i as the expected number of times a transition from s_i will take place:

$$\tau_i = \sum_{k=1}^{N-1} \gamma_k(i) \qquad (3.94)$$

This quantity will be computed by the MATLAB® code described in Listing 3.6.

Listing 3.6 MATLAB® function for computing τ_i.

```
1  function taui=compute_taui(Gama,B,O)
2  % Compute nu ..
3  % m hidden states, n output states and N observations
4  %
5  % Gama - Nxm (from the forward algorithm)
6  % O - 1xN (observations vector)
7  % nu - Return an mxm matrix
8
9  [m,~]=size(B);
```

```
10  N=length(O);
11  taui=Gama(1:N-1,:);
12  taui=(sum(taui,1)).'*ones(1,m);
```

Furthermore, note the close relation between $\gamma_k(i)$ and $\xi_k(i,j)$. As a matter of fact,

$$\gamma_k(i) = \sum_{j=1}^{m} \xi_k(i,j) \qquad (3.95)$$

This expression can be interpreted as follows: the state s_i activation probability at time k is equal to the sum of the transition probabilities from state s_i to any of the possible m hidden states at time $k+1$.

It lacks to find an appropriate strategy to determine $\gamma_k(i)$ and $\xi_k(i,j)$. Starting with equation (3.90) and multiplying both terms by $P(\Lambda_N)$ leads to:

$$\gamma_k(i) \cdot P(\Lambda_N) = P(q_k = s_i | \Lambda_N) P(\Lambda_N) \qquad (3.96)$$

Resorting to the Bayes theorem, the above expression can be rewritten as:

$$\gamma_k(i) \cdot P(\Lambda_N) = P(\Lambda_N | q_k = s_i) \cdot P(q_k = s_i) \qquad (3.97)$$

Since it is assumed statistical independence between observations then:

$$\gamma_k(i) \cdot P(\Lambda_N) = P(\Lambda_k | q_k = s_i) \cdot P(\Lambda_{k+1 \to N} | q_k = s_i) \cdot P(q_k = s_i)$$
$$= P(\lambda_1 \cdots \lambda_k | q_k = s_i) \cdot P(\lambda_{k+1} \cdots \lambda_N | q_k = s_i) \cdot P(q_k = s_i) \qquad (3.98)$$

As was seen in Section 3.2.2, given the state $q_k = s_i$, the probability of observing $\lambda_{k+1} \cdots \lambda_N$ is $\beta_k(i)$. Consequently, after replacing $P(\lambda_{k+1} \cdots \lambda_N | q_k = s_i)$ by $\beta_k(i)$, the above expression takes the following form:

$$\gamma_k(i) \cdot P(\Lambda_N) = P(\lambda_1 \cdots \lambda_k | q_k = s_i) \cdot P(q_k = s_i) \cdot \beta_k(i) \qquad (3.99)$$

From the Bayes theorem it is known that,

$$P(\lambda_1 \cdots \lambda_k | q_k = s_i) \cdot P(q_k = s_i) = P(\lambda_1 \cdots \lambda_k \wedge q_k = s_i) \qquad (3.100)$$

then,

$$\gamma_k(i) \cdot P(\Lambda_N) = P(\lambda_1 \cdots \lambda_k \wedge q_k = s_i) \cdot \beta_k(i) \qquad (3.101)$$

Furthermore, taking into consideration that,

$$\alpha_k(i) = P(\lambda_1 \cdots \lambda_k \wedge q_k = s_i) \qquad (3.102)$$

equation (3.99) becomes,

$$\gamma_k(i) = \frac{\alpha_k(i)\,\beta_k(i)}{P(\Lambda_N)} \qquad (3.103)$$

where the denominator can be obtained by $P(\Lambda_N) = \sum_{i=1}^{m} \alpha_N(i)$. Alternatively, since $\alpha_N(i) = \alpha_k(i)\,\beta_k(i)$, the probability of having s_i as the active hidden state at time k, given the information provided by Λ_N is:

$$\gamma_k(i) = \frac{\alpha_k(i)\,\beta_k(i)}{\sum_{j=1}^{m} \alpha_k(j)\,\beta_k(j)} \qquad (3.104)$$

The MATLAB® function presented in Listing 3.7 will be used to compute this probability.

Listing 3.7 MATLAB® function for computing γ.

```
1  function Gama=compute_gama(Alfa,Beta)
2  % Compute gamma
3  % Alfa - Nxm (from the forward algorithm)
4  % Beta - Nxm (from the backward algorithm)
5  % Gama - Return an Nxm matrix with the shape:
6  %        _                              _
7  % |   gama_1(1)  ...  gama_1(m)  |
8  % |      ...              ...       |
9  % |   gama_N(1)  ...  gama_N(m)  |
10 %        -                              -
11
12 [~,m]=size(Alfa);
13 P=diag(Alfa*Beta')*ones(1,m);
14 Gama=(Alfa.*Beta)./P;
```

An interesting feature of the above expression is that it can be computed efficiently using the forward and backward algorithms described previously.

Using $\gamma_k(i)$ it is even possible to find which of the m states is more likely to be active at time k. Indeed, the state whose $\gamma_k(i)$ is larger, relatively to the remains, will be the most likely. That is, the more likely active hidden state at time k is $q_k = s_i$ where $i = \arg\max_p \gamma_k(p)$.

The transition probability from hidden state s_i to s_j, provided the set of observations Λ_N, can also be efficiently computed by means of the forward and backward algorithms. In particular, consider equation (3.92). Multiplying both terms by $P(\Lambda_N)$, and applying Bayes theorem, leads to,

$$\xi_k(i,j) \cdot P(\Lambda_N) = P(q_k = s_i \wedge q_{k+1} = s_j \wedge \Lambda_N) \tag{3.105}$$

The observations sequence can be partitioned as $\Lambda_N = \{\Lambda_k \ \Lambda_{k+1\to N}\}$ where $\Lambda_{k+1\to N}$ regards the observations from time $k+1$ to N. That is, $\Lambda_{k+1\to N} = \{\lambda_{k+1} \cdots \lambda_N\}$. Since the observations are statistically independent one from each other then,

$$\begin{aligned}
&P(q_k = s_i \wedge q_{k+1} = s_j \wedge \Lambda_N) = \\
&P(\Lambda_k \wedge q_{k+1} = s_j \wedge q_k = s_i) \cdot P(\Lambda_{k+1\to N} | q_{k+1} = s_j \wedge q_k = s_i)
\end{aligned} \tag{3.106}$$

Replacing the above expression into (3.105), multiplying and dividing by $P(q_{k+1} = s_j \wedge q_k = s_i)$ results in:

$$\begin{aligned}
&\xi_k(i,j) \cdot P(\Lambda_N) = \\
&\frac{P(\Lambda_k \wedge q_{k+1} = s_j \wedge q_k = s_i)}{P(q_{k+1} = s_j \wedge q_k = s_i)} \cdot P(q_{k+1} = s_j \wedge q_k = s_i) \cdot \\
&P(\Lambda_{k+1\to N} | q_{k+1} = s_j \wedge q_k = s_i)
\end{aligned} \tag{3.107}$$

due to the fact that,

$$\frac{P(\Lambda_k \wedge q_{k+1} = s_j \wedge q_k = s_i)}{P(q_{k+1} = s_j \wedge q_k = s_i)} = P(\Lambda_k | q_{k+1} = s_j \wedge q_k = s_i) \tag{3.108}$$

and

$$\begin{aligned}
&P(q_{k+1} = s_j \wedge q_k = s_i) \, P(\Lambda_{k+1\to N} | q_{k+1} = s_j \wedge q_k = s_i) \\
&= P(\Lambda_{k+1\to N} \wedge q_{k+1} = s_j \wedge q_k = s_i)
\end{aligned} \tag{3.109}$$

then,

$$\begin{aligned}
&\xi_k(i,j) \cdot P(\Lambda_N) = \\
&P(\Lambda_k | q_{k+1} = s_j \wedge q_k = s_i) \cdot P(\Lambda_{k+1\to N} \wedge q_{k+1} = s_j \wedge q_k = s_i)
\end{aligned} \tag{3.110}$$

Since Λ_k only depends on the active hidden state at time instant k, i.e.,

$$P(\Lambda_k | q_{k+1} = s_j \wedge q_k = s_i) = P(\Lambda_k | q_k = s_i) \tag{3.111}$$

then,

$$\begin{aligned}
&\xi_k(i,j) \cdot P(\Lambda_N) = \\
&P(\Lambda_k | q_k = s_i) \cdot P(\Lambda_{k+1\to N} \wedge q_{k+1} = s_j \wedge q_k = s_i)
\end{aligned} \tag{3.112}$$

Taking Bayes theorem into consideration, the above expression is now described as:

$$\xi_k(i,j) \cdot P(\Lambda_N) =$$
$$P(\Lambda_k \wedge q_k = s_i) \frac{P(\Lambda_{k+1\to N} \wedge q_{k+1} = s_j \wedge q_k = s_i)}{P(q_k = s_i)} \qquad (3.113)$$

Since $\Lambda_{k+1\to N}$ is independent from q_k,

$$\frac{P(\Lambda_{k+1\to N} \wedge q_{k+1} = s_j \wedge q_k = s_i)}{P(q_k = s_i)} =$$
$$P(\Lambda_{k+1\to N} | q_{k+1} = s_j) \cdot P(q_{k+1} = s_j | q_k = s_i) \qquad (3.114)$$

the expression (3.113) becomes,

$$\xi_k(i,j) \cdot P(\Lambda_N) =$$
$$P(\Lambda_k \wedge q_k = s_i) \cdot P(\Lambda_{k+1\to N} | q_{k+1} = s_j) \cdot P(q_{k+1} = s_j | q_k = s_i) \qquad (3.115)$$

Notice that $\Lambda_{k+1\to N} = \lambda_{k+1} \wedge \Lambda_{k+2\to N}$. For this reason,

$$P(\Lambda_{k+1\to N} | q_{k+1} = s_j) = P(\lambda_{k+1} \wedge \Lambda_{k+2\to N} | q_{k+1} = s_j) \qquad (3.116)$$

Due to the fact that the observations are statistically independent[3], the following expression is obtained:

$$P(\Lambda_{k+1\to N} | q_{k+1} = s_j) =$$
$$P(\lambda_{k+1} | q_{k+1} = s_j) \cdot P(\Lambda_{k+2\to N} | q_{k+1} = s_j) \qquad (3.117)$$

Next, according to this last equality, (3.115) becomes:

$$\xi_k(i,j) \cdot P(\Lambda_N) = P(\Lambda_k \wedge q_k = s_i) \cdot P(\lambda_{k+1} | q_{k+1} = s_j) \cdot$$
$$P(\Lambda_{k+2\to N} | q_{k+1} = s_j) \cdot P(q_{k+1} = s_j | q_k = s_i) \qquad (3.118)$$

Taking into account that,

$$P(\Lambda_k \wedge q_k = s_i) = \alpha_k(i)$$
$$P(q_{k+1} = s_j | q_k = s_i) = a_{ij}$$
$$P(\lambda_{k+1} | q_{k+1} = s_j) = b_j(\lambda_{k+1})$$
$$P(\Lambda_{k+2\to N} | q_{k+1} = s_j) = \beta_{k+1}(j) \qquad (3.119)$$

---

[3] $\lambda_{k+1} \perp \Lambda_{k+2\to N}$

equation (3.118) is rewritten as:

$$\xi_k\left(i,j\right)\cdot P\left(\Lambda_N\right)=\alpha_k\left(i\right)a_{ij}b_j\left(\lambda_{k+1}\right)\beta_{k+1}\left(j\right) \tag{3.120}$$

Leading, finally, to:

$$\begin{aligned}\xi_k\left(i,j\right)&=\frac{\alpha_k\left(i\right)a_{ij}b_j\left(\lambda_{k+1}\right)\beta_{k+1}\left(j\right)}{P\left(\Lambda_N\right)}\\&=\frac{\alpha_k\left(i\right)a_{ij}b_j\left(\lambda_{k+1}\right)\beta_{k+1}\left(j\right)}{\sum_{i=1}^m\sum_{j=1}^m\alpha_k\left(i\right)a_{ij}b_j\left(\lambda_{k+1}\right)\beta_{k+1}\left(j\right)}\end{aligned} \tag{3.121}$$

It is worth noting that, as with $\gamma_k\left(i\right)$, this last expression only depends on factors that can be efficiently calculated by both the forward and backward algorithms.

At this point, we possess the ability to efficiently compute the expected number of times that the state s_i is visited (denoted by ν_i), the expected number of transitions leaving the state s_i (defined by τ_i) and the expected number of transitions from state s_i to s_j (that is, τ_{ij}). Based on these values, the Baum-Welch training algorithm will iteratively estimate the model parameters Θ resorting only to a set of observations [37]. In particular, the Baum-Welch method computes an estimation of c_i and a_{ij}, denoted here by $\hat{c}_i$ and $\hat{a}_{ij}$, using:

$$\hat{c}_i=\gamma_1\left(i\right),\quad1\le i\le m \tag{3.122}$$

$$\hat{a}_{ij}=\frac{\tau_{ij}}{\tau_i}=\frac{\sum_{k=1}^{N-1}\xi_k\left(i,j\right)}{\sum_{k=1}^{N-1}\gamma_k\left(i\right)} \tag{3.123}$$

The estimation formula for c_i is simply the model probability to be in the state s_i at time $k=1$. On the other hand, the estimation of a_{ij} requires the computation of the ratio between the expected number of transitions from state s_i to the state s_j and the total number of state transitions leaving s_i.

In the same way, estimation for b_{ij}, hereafter denoted by $\hat{b}_{ij}$, is computed as the ratio between the number of transitions expected value for the observable state r_j, for $j=1,\cdots,n$, given the hidden state s_i, and the total number of visits to state s_i. According to this definition, the calculation of $\hat{b}_{ij}$ requires the knowledge of the number of times that, being at state s_i, the state r_j is observed. This value, denoted by ω_{ij}, is determined as:

$$\omega_{ij}=\sum_{k=1}^N\gamma_k\left(i\right)\cdot\delta\left(\lambda_k-r_j\right) \tag{3.124}$$

where, in this context, δ refers to the Kronecker delta function. That is, $\delta(\lambda_k - r_j)$ is equal to 1 if, and only if, $\lambda_k = r_j$. Otherwise it is zero. The MATLAB® code presented in Listing 3.8 will be used to compute ω_{ij}.

Listing 3.8 MATLAB® function for computing ω_{ij}.

```
 1  function Omega=compute_omega(Gama,B,O)
 2  % Compute omega...
 3  % m hidden states, n output states and N observations
 4  % Gama - Nxm matrix
 5  % B - nxm (confusion matrix)
 6  % O - 1xN (observations vector)
 7
 8  [m,n]=size(B);
 9  for j=1:n,
10      inx=find(O==j);
11      if ~isempty(inx),
12          Omega(:,j)=sum(Gama(inx,:),1).';
13      else
14          Omega(:,j)=0*ones(m,1);
15      end
16  end
```

Hence, the value for b_{ij} is estimated according to:

$$\hat{b}_{ij} = \frac{\omega_{ij}}{\nu_i} = \frac{\sum_{k=1}^{N} \gamma_k(i) \cdot \delta(\Lambda_k - r_j)}{\sum_{k=1}^{N} \gamma_k(i)} \tag{3.125}$$

which is nothing more than the ratio between the number of times that, being in state s_i, r_j was observed and the expected number of times the model will be in state s_i.

The set of probabilities estimation equations for $\hat{a}_{ij}$, $\hat{b}_{ij}$ and $\hat{c}_i$, for $i = 1, \cdots, m$ and $j = 1, \cdots n$ are not in closed form. Hence, they are aggregated into an iterative procedure denoted by Baum-Welch estimation method. Additional insights on this procedure can be obtained by analysing the MATLAB® code presented in Listing 3.9.

Listing 3.9 Baum-Welch algorithm coded for MATLAB®.

```
 1  function [A,B,c]=baum_welch(A,B,O,c)
 2  % Baum-Welch algorithm
 3  % m hidden states, n output states and N observations
 4  % A - mxm (state transitions matrix)
 5  % B - nxm (confusion matrix)
 6  % O - 1xN (observations vector)
 7  % c - 1xm (priors vector)
 8
 9  Alfa  = forward_algorithm(A,B,O,c);
```

```
10 Beta  = backward_algorithm(A,B,O);
11 Gama  = compute_gama(Alfa,Beta);
12 tau   = compute_tau(Alfa,Beta,A,B,O);
13 taui  = compute_taui(Gama,B,O);
14 nu    = compute_nu(Gama,B);
15 Omega = compute_omega(Gama,B,O);
16 c     = Gama(1,:);
17 A     = tau./taui;
18 B     = Omega./nu;
```

This computer code returns a new estimation for matrices $\mathbf{A}$, $\mathbf{B}$ and vector $\mathbf{c}$. Observe the call to both the backward and forward functions described on Sections 3.2.1 and 3.2.2.

Just before ending this section, it is important to underline that there is convergence proof for this algorithm [38]. The model likelihood, after iteratively executing equations (3.122), (3.123) and (3.125), will improve or maintain its value. That is, at iteration $l + 1$, the model likelihood obtained by using the estimated parameters $\hat{\Theta}_{l+1}$ will not be lower than the likelihood measured with the previous estimated parameters. Strictly speaking,

$$\mathcal{L}\left(\hat{\Theta}_l \middle| \Lambda_N\right) \leq \mathcal{L}\left(\hat{\Theta}_{l+1} \middle| \Lambda_N\right) \tag{3.126}$$

This condition implies that this algorithm will converge toward a maximum. However, due to the non-convexity problem, there are no guarantees that the global optimum will be attained. Moreover, the solution presented by the algorithm, at the end of all iterations, will depend heavily on the initial parameters estimation.

3.3.2.1 Operation conditions for the Baum-Welch algorithm

The probability of an event is obtained, *a posteriori*, conducting a theoretically infinite number of experiences. Its value is attained by taking the ratio between the number of times the desired event was observed and the total number of experiments conduced. Since it is impossible to have an infinite number of experiments, its value must be truncated somewhere. Care must be taken since, if the number of experiments is too small, the estimated probability will be biased. For example, if the outcome of flipping a regular coin twice is "heads" then, computing the *a posteriori* probability for "heads", will be equal to one. A value known to be false for an unbiased coin.

For this reason, hidden Markov model parameter estimation, by means of the Baum-Welch method, can lead to wrong probability estimations if the training set dimension is inadequate. Imagine a situation where there is a state that it is not used by the training sequence set. In this case, the estimation equations lead to undefined values because the numerator of both equations (3.123) and (3.125) is zero.

The data can therefore be insufficient to excite all directions of the parameter space. In these situations, if possible, one should increase the number of training observations. If this strategy is not feasible, then reducing the model dimensionality should be equated. In some cases, the estimation data set is expanded by adding pseudo-counts. These pseudo-counts are artificial observations and should reflect some prior knowledge about the model [23].

Another problem results from the objective function non-convexity associated to the optimization problem. For this reason, the model performance strongly depends on the initial guessed parameters location. Usually this initial solution can be generated randomly or, alternatively, by using uniform random values for **A** and **c** and some segmentation strategy for **B**. This segmentation can be carried out by using averages, k-means or any other data clustering technique.

3.3.2.2 Parameter estimation using multiple trials

Let us assume the situation where a set of L experimental trials is carried out. For each trial, a set of N observations is registered. On the other hand, the same number of $L \times N$ observations could be obtained in a single experiment. Is there any difference regarding the model parameters estimation in using the former or the latter data set?

First of all, the use of shorter sequences minimizes the effect of numerical algorithm instability. The explanation for this fact will be provided in subsection 3.3.2.3. At the same time, each Baum-Welch algorithm iterative cycle requires the computation of partial $\gamma_k(i)$, τ_i, τ_{ij}, ω_{ij} and ν_i for each one of the L training sequences. Let $\gamma_k^l(i)$, τ_i^l, τ_{ij}^l, ω_{ij}^l and ν_i^l be those partial quantities for $l = 1, \cdots, L$. For each algorithm iteration, the overall values for variables τ_i, τ_{ij}, ω_{ij} and ν_i are computed by summing τ_i^l, τ_{ij}^l, ω_{ij}^l and ν_i^l across all the L data sets. That is,

$$\gamma_k(i) = \sum_{l=1}^{L} \gamma_k^l(i) \quad \tau_{ij} = \sum_{l=1}^{L} \tau_{ij}^l \quad \tau_i = \sum_{l=1}^{L} \tau_i^l$$
$$\nu_i = \sum_{l=1}^{L} \nu_i^l \quad \omega_{ij} = \sum_{l=1}^{L} \omega_{ij}^l \quad (3.127)$$

Due to observations' statistical independence, the overall likelihood value is computed by taking the likelihood product of each one of the L sequences. That is,

$$\mathcal{L}(\Theta|\Lambda_{N\times L}) = \prod_{l=1}^{L} \mathcal{L}(\Theta|\Lambda_N^l) \tag{3.128}$$

In order to minimize the underflow problem, due to the product of successive terms smaller than one, the likelihood logarithm is computed instead of the raw likelihood. This approach leads to:

$$\mathcal{L}^e(\Theta|\Lambda_{N\times L}) = \sum_{l=1}^{L} \log\left(\mathcal{L}(\Theta|\Lambda_N^l)\right) \tag{3.129}$$

The problem of numerical stability due to the successive multiplication of several elements with values less than unity exists also in both the forward and backward algorithms. For this reason, and in order to increase the numerical stability, the following section is devoted to the modification of the original forward and backward algorithms.

3.3.2.3 Baum-Welch algorithm numerical stability

As was seen, the parameters computation of a hidden Markov model using the Baum-Welch algorithm requires the calculation of the partial probabilities $\alpha_k(i)$. Since these values demand the estimation of joint probabilities throughout the sequence up to the time k, their values tend to be very small. This comes from the fact that the multiplication of successive values, lower than unity, is observed. In this framework, even for medium size observations sequences, the numeric resolution of an ordinary computer is easily exceeded.

In order to circumvent this problem, [23] proposes the addition of scale factors to normalize the values of α_k and β_k. Those factors must be independent of the active hidden state but should depend on k.

Let $\bar{\alpha}_k$ and $\bar{\beta}_k$ be the scaled versions of α_k and β_k. At each time instant k, the forward and backward probabilities are multiplied by a weight factor. That operation coerces their values to be within the computer numerical representation capability. That is, for $k = 1$, the scaled forward probability $\bar{\alpha}_1$ is computed by $\bar{\alpha}_1 = p_1 \cdot \alpha_1$. For $k = 2$ by $\bar{\alpha}_2 = p_1 \cdot p_2 \cdot \alpha_2$ and so on. The parameters p_i are the scale coefficients.

In short, for an arbitrary time instant k, the scaled forward probability is computed by:

$$\bar{\alpha}_k(i) = p_1 \cdot \ldots \cdot p_k \cdot \alpha_k(i) = u_k \cdot \alpha_k(i) \qquad (3.130)$$

where,

$$u_k = \prod_{j=1}^{k} p_j \qquad (3.131)$$

The values for u_k can be arbitrarily chosen. However, they usually are selected, at time instant k, in order to normalize to unity the value of $\bar{\alpha}_k$. That is, the value for u_k will be computed by:

$$u_k = \frac{1}{\sum_{i=1}^{m} \alpha_k(i)} \qquad (3.132)$$

Hence, for each time instant k, the weight factor is inversely proportional to the probability $P(\Lambda_k)$. The MATLAB® code presented in Listing 3.10 describes the normalised version of the forward algorithm.

Listing 3.10 Normalised forward algorithm coded for MATLAB®.

```
1  function [Alfa,LogLik]=forward_algorithm_norm(A,B,O,c)
2  % Forward algorithm with normalization
3  %[Alfa,LogLik]=forward_algorithm_norm(A,B,O,c)
4  % m hidden states, n output states and N observations
5  % A - mxm (state transitions matrix)
6  % B - nxm (confusion matrix)
7  % O - 1xN (observations vector)
8  % c - 1xm (priors vector)
9  %
10 % Alfa - Partial probability matrix P(Lambda_k ^ q_k).
11 % LogLik - log likelihood of the observation sequence O
12
13
14 [m,~]=size(B);
15 N=length(O);
16 u=zeros(1,N);
17
18 %% Initialization
19 Alfa=zeros(N,m);
20 for k=1:m
21   Alfa(1,k)=c(k)*B(k,O(1));
22 end
23 u(1)=1/sum(Alfa(1,:));        % Scaling coefficient
24 Alfa(1,:)=u(1)*Alfa(1,:);
25
26 %% Recursion
27 for l=2:N,
28   for k=1:m,
29     S=0;
```

```
30      for i=1:m,
31          S=S+A(i,k)*Alfa(1-1,i);
32      end
33      Alfa(1,k)=B(k,O(1))*S;
34   end
35   u(1)=1/(sum(Alfa(1,:))); % Scaling coefficient
36   Alfa(1,:)=u(1)*Alfa(1,:);
37 end
38
39 %% Compute Log Likelihood
40 LogLik=-sum(log(u));
```

Regarding the backward probability $\beta_k(i)$, Rabiner in [23] suggests to use as scaling factors the same u_k parameters computed for the forward probability. That is,

$$\bar{\beta}_k(i) = u_k \cdot \beta_k(i) \tag{3.133}$$

This scaling coefficients choice is justified by the need to promote factorizations ahead in the normalised Baum-Welch equations. However, as one will see in short, any arbitrary value could be selected. The normalised version of the Baum-Welch equations will not suffer if a distinct value was used. In this book, and only to be coherent with the constraints imposed to α_k, the scaling factors for $\bar{\beta}_k$, denoted hereafter by ν_k, are computed by:

$$v_k = \frac{1}{\sum_{i=1}^{m} \beta_k(i)} \tag{3.134}$$

It is worth noting that, with this choice for $\bar{\beta}_k$ scaling coefficients, the condition $\sum_{i=1}^{m} \bar{\beta}_k(i) = 1$ will be met. This is similar to the condition associated to $\bar{\alpha}_k$. However, due to the backward computation nature of β_k, the value of $\bar{\beta}_k$, for an arbitrary state i, is given by:

$$\bar{\beta}_k(i) = v_k \cdot \beta_k(i) \tag{3.135}$$

Due to this transformation, the backward algorithm presented earlier, is slightly changed to encompass the normalising factor. The new MATLAB® version is presented in Listing 3.11.

Listing 3.11 Normalised backward algorithm coded for MATLAB®.

```
1 function Beta=backward_algorithm_norm(A,B,O)
2 % Backward algorithm with normalization
3 %[Alfa,LogLik]=backward_algorithm(A,B,O,c)
4 % m hidden states, n output states and N observations
5 % A - mxm (state transitions matrix)
6 % B - nxm (confusion matrix)
```

```
 7  % O - 1xN (observations vector)
 8
 9  [m,~]=size(B);
10  N=length(O);
11
12  %% Initialization
13  Beta=zeros(N,m);
14  for k=1:m
15    Beta(N,k)=1;
16  end
17  v(N)=1/sum(Beta(N,:));        % Scaling coefficient
18
19  %% Recursion
20  for t=N-1:-1:1,
21    for i=1:m,
22      Beta(t,i)=0;
23      for j=1:m,
24        Beta(t,i)=Beta(t,i)+A(i,j)*B(j,O(t+1))*Beta(t+1,j);
25      end
26    end
27    v(t)=1/sum(Beta(t,:));  % Scaling coefficient
28    Beta(t,:)=v(t)*Beta(t,:);
29  end
```

This normalization process will not change any of the Baum-Welch equations. Making the replacement $\alpha_k(i) \to \bar{\alpha}_k(i)$ and $\beta_k(i) \to \bar{\beta}_k(i)$ at equations (3.104) and (3.89) will lead, after some algebraic operations, to $\bar{\gamma}_k = \gamma_k(i)$ and $\bar{\xi}_k(i) = \xi_k(i)$. However, after replacing α_k by $\bar{\alpha}_k$, the observations likelihood cannot be computed using (3.89). Instead, after simple algebraic manipulation of (3.89) and by taking into consideration (3.130), the likelihood should be computed using:

$$\mathcal{L}(\Theta|\Lambda_k) = \frac{1}{\prod_{i=1}^{k} u_i} \tag{3.136}$$

or, alternatively, the log-likelihood version computed as:

$$\mathcal{L}^e(\Theta|\Lambda_k) = -\sum_{j=1}^{k} \ln(u_j) \tag{3.137}$$

This likelihood computation strategy is considered in Listing 3.10 and the normalised version of the Baum-Welch method using "multi sets" is presented in Listing 3.12.

Listing 3.12 The "multi sets" version of the Baum-Welch algorithm coded for MATLAB®.

```matlab
1  function [A,B,c,LogLik]=baum_welch_multiobs_norm(A,B,O,c)
2  % Baum-Welch algorithm: for multiobservations and with normalization
3  % [A,B,c,LogLik]=baum_welch_multiobs(A,B,O,c)
4  % m hidden states, n output states and N observations
5  % A  - mxm (state transitions matrix)
6  % B  - nxm (confusion matrix)
7  % O  - 1xN (observations vector)
8  % c  - 1xm (priors vector)
9  % LogLik - log likelihood of the observation sequence O
10
11 [route_nbr,~]=size(O);
12 GamaT=0;
13 tauT=0;
14 tauiT=0;
15 OmegaT=0;
16 nuT=0;
17 LogLik=0;
18
19 for route=1:route_nbr,
20    [Alfa,LP]=forward_algorithm_norm(A,B,O(route,:),c);
21    LogLik=LogLik+LP;
22    Beta=backward_algorithm_norm(A,B,O(route,:));
23    Gama=compute_gama(Alfa,Beta);
24    GamaT=GamaT+Gama;
25    tau=compute_tau(Alfa,Beta,A,B,O(route,:));
26    tauT=tauT+tau;
27    taui=compute_taui(Gama,B,O(route,:));
28    tauiT=tauiT+taui;
29    nu=compute_nu(Gama,B);
30    nuT=nuT+nu;
31    Omega=compute_omega(Gama,B,O(route,:));
32    OmegaT=OmegaT+Omega;
33 end
34
35 c=GamaT(1,:)/route_nbr;
36 A=tauT./tauiT;
37 B=OmegaT./nuT;
```

The Baum-Welch algorithm is, by no means, the only way to estimate the values for the transition probabilities of a hidden Markov model. The next two sections will cover other possible parameters estimation strategies.

3.4 VITERBI TRAINING ALGORITHM

In the hidden Markov model training methods that involve the improvement of the maximum likelihood, such as the Baum-Welch algorithm, the goal is to determine a set of Θ parameters that maximizes $\mathcal{L}(\Theta|\Lambda_N)$ for a sequence of N observations $\Lambda_N = \{\lambda_1, \cdots, \lambda_N\}$. However, the algorithm described in this section, which was proposed by [13], attempts to determine the Markov model by maximizing an alternative figure-of-merit. In this method, the probability of the observations sequence is not obtained taking into account all possible paths, as happens in maximum likelihood, but only taking into consideration the *most likely* path. On other words, it is intended to maximize[4] $P(\Lambda|Q^* \wedge Q^*)$ where Q^* refers to the most probable sequence of hidden states, $\{q_0^* q_1^* \cdots q_N^*\}$, assuming the knowledge of the observable states sequence Λ_N. As previously seen, this sequence of states, after known the parameters of the Θ model, can be determined using the Viterbi algorithm. The benchmark that is now shown to be maximized was already considered before in (3.21) and shown again below.

$$P\left(\Lambda|Q^* \wedge Q^*\right) = c_{\delta_1} b_{\delta_1}\left(\lambda_1\right) \cdot \prod_{k=2}^{N} a_{\delta_{k-1}\delta_k} b_{\delta_k}\left(\lambda_k\right) \tag{3.138}$$

where, again, δ_k refers to the state index pointed to by q_k^*. Note that the computational load involved in determining the previous expression is much smaller than that required to calculate the likelihood function.

The method designated herein by "Viterbi training" involves two distinct phases. In a first step, the optimum sequence Q^* is obtained by taking into account both the training data and a particular model with parameters Θ. The next step involves the computation of the new model parameters using a counting method similar to the Baum-Welch method. However, in this case, the counting is taken assuming only the sequence Q^*. The equations for updating the parameters are:

$$\hat{a}_{ij} = \frac{\sum_{k=1}^{N-1} \delta(q_k^* - s_i)\delta(q_{k+1} - s_j)}{\sum_{k=1}^{N-1} \delta(q_k^* - s_i)} \tag{3.139}$$

$$\hat{b}_i(\lambda_j) = \frac{\sum_{k:\Lambda_k=\lambda_j} \delta(q_k^* - s_i)}{\sum_{k=1}^{N-1} \delta(q_k^* - s_i)} \tag{3.140}$$

where $\delta(\cdot)$ refers to the Kronecker function defined as,

$$\delta(q_k^* - s_i) = \begin{cases} 1 & \text{if} \quad q_k^* = s_i \\ 0 & \text{otherwise} \end{cases} \tag{3.141}$$

and,

$$\hat{c}_i = \delta(q_1^* - s_i) \tag{3.142}$$

---

[4]For reasons related to the lightness of notation, the explicit dependence of the probability value with Θ is waived.

It is important to highlight some remarks about these equations. First of all, the indicator function $\delta(q_k^* - s_i)$ returns the value 1 if the active state, for the observation k, coincides with the most probable state for the same observation. For example, suppose that the sequence of states obtained by the Viterbi algorithm was $Q^* = \{s_3, s_1, s_1, s_2\}$. In this case, $\delta(q_2^* - s_2) = 0$ and $\delta(q_3^* - s_1) = 1$. The denominator of both (3.139) and (3.140), corresponds to the number of times that the state s_i appears in the Viterbi sequence.

The numerator of the first expression involves the sum of the product of two indicator functions: one that depends on the present state i and another that depends on the next state j. Notice that the product $\delta(q_k^* - s_i)\delta(q_{k+1}^* - s_j)$ is only different from zero if, for the observation k, the hidden state s_i matches q_k^* and the hidden state s_j matches q_{k+1}^*.

The numerator of the equation (3.140) involves counting the number of times where in $s_i = q_k^*$, the symbol λ_j is observed.

The pseudo-code represented in Algorithm 1 summarizes the steps to be executed in order to perform the re-estimation of the model parameters by the Viterbi training method. One last note about the initial probability

---

Algorithm 1 Viterbi training algorithm.

---

1. An initial $\hat{\Theta}_0$ is defined (for example randomly).
2. Determine for the $\hat{\Theta}_0$ model and based on a training sequence Λ with N examples, the most probable sequence Q^* based on the Viterbi algorithm.
3. Re-estimate model parameters:

$$\hat{a}_{ij} = \frac{\sum_{k=1}^{N-1} \delta(q_k^* - s_i)\delta(q_{k+1}^* - s_j)}{\sum_{k=1}^{N-1} \delta(q_k^* - s_i)}$$

$$\hat{b}_i(\lambda_j) = \frac{\sum_{k:\Lambda_k = \lambda_j} \delta(q_k^* - s_i)}{\sum_{k=1}^{N-1} \delta(q_k^* - s_i)}$$

$$\hat{c}_i = \delta(q_1^* - s_i)$$

where,

$$\delta(q_k^* - s_i) = \begin{cases} 1 & \text{if } q_k^* = s_i \\ 0 & \text{otherwise} \end{cases}$$

4. Repeat step (2) until some stop criterion is satisfied.

---

vector **c**. Since only a single first observation exists in the sequence, the unity probability is provided to the state associated with the one resulting from the Viterbi algorithm.

Finally, the Viterbi training method, coded in MATLAB®, is presented in Listing 3.13. Additional information regarding the Viterbi training method can be found, for example, in [16].

Listing 3.13 Coding of Viterbi training algorithm.

```
1  function [A,B,c,Fit]=viterbi_learning(m,n,0,MaxIter)
2  % Viterbi learning algorithm
3  % m hidden states, n output states and N observations
4  % O - 1xN (observations vector)
5  % MaxIter - maximum number of iterations
6
7  N=length(0);
8
9  % Random PARAMETERS INITIALIZATION
10 A=rand(m,m);A=A./(sum(A').'*ones(1,2));
11 B=rand(m,n);B=B./(sum(B').'*ones(1,3));
12 c=rand(m,1);c=c/sum(c);
13
14 % ITERATION
15 for k=1:MaxIter,
16
17   % Determines performance of new parameters...
18   [pd,P]=forward_algorithm_norm(A,B,0,c);
19   disp(['Iteration -- ' num2str(k) ' FIT -- ' num2str(P)]);
20   Fit(k)=P;
21
22   % SEGMENTATION: VITERBI ALGORITHM
23   Q=viterbi_algorithm(A,B,0,c).';
24
25   % ESTIMATION c
26   c=zeros(2,1);c(Q(1))=1;
27
28   % ESTIMATION A
29   for i=1:m,
30     for j=1:m,
31       Num=(sum(mu(i*ones(1,N-1),Q(1:N-1)).*mu(j*ones(1,N-1),Q(2:N))));
32       Den=(sum(mu(i*ones(1,N-1),Q(1:N-1))));
33       if Den~=0,
34         A(i,j)=Num/Den;
35       else
36         A(i,j)=1;
37       end
38     end
39   end
40
41   % ESTIMATION B
42   for i=1:m,
43     for j=1:n,
44       k=find(0==j);
45       if isempty(k),
46         B(i,j)=0;
47       else
48         Num=(sum(mu(i*ones(1,length(k)),Q(k))));
49         Den=(sum(mu(i*ones(1,N),Q(1:N))));
50         if Den~=0,
51           B(i,j)=Num/Den;
52         else
53           B(i,j)=1;
54         end
55       end
56     end
57   end
58 end
```

81

3.5 GRADIENT-BASED ALGORITHMS

The most common method for hidden Markov model parameters estimation is, without any doubt, the Baum-Welch method. However, the maximization of the likelihood function can be also achieved by methods involving the gradient (and Hessian) of the log likelihood function [26]. For example, the gradient ascent algorithms or even second derivative approaches such as the Levenberg-Marquardt optimization method.

The reason underlying the use of these tools lies in the fact that the likelihood function $\mathcal{L}$ is differentiable along the model parameters $\boldsymbol{\Theta}$. Using the information of $\nabla \mathcal{L}$, around an arbitrary point in the parameter space $\hat{\boldsymbol{\Theta}}_l$, allows us to estimate a new vector $\hat{\boldsymbol{\Theta}}_{l+1}$ such that:

$$\mathcal{L}\left(\hat{\boldsymbol{\Theta}}_l | \Lambda_k\right) < \mathcal{L}\left(\hat{\boldsymbol{\Theta}}_{l+1} | \Lambda_k\right) \qquad (3.143)$$

One of the possibilities for calculating $\hat{\boldsymbol{\Theta}}_{l+1}$ from $\hat{\boldsymbol{\Theta}}_l$ consists on the famous gradient ascent algorithm. In this technique, the parameters of the model are updated, iteratively according to the law:

$$\hat{\boldsymbol{\Theta}}_{l+1} = \hat{\boldsymbol{\Theta}}_l + \eta \cdot \nabla \mathcal{L}_k |_{\boldsymbol{\Theta} = \hat{\boldsymbol{\Theta}}_l} \qquad (3.144)$$

where η refers to a positive parameter denoted by learning ratio, l refers to the iteration index, and $\nabla \mathcal{L}_k$ refers to the likelihood gradient, along $\boldsymbol{\Theta}$, calculated for the first k observations of Λ_N.

3.5.1 PARTIAL DERIVATIVE OF $\mathcal{L}_K$

Since $\boldsymbol{\Theta} \to (\mathbf{A}, \mathbf{B}, \mathbf{c})$, then $\nabla \mathcal{L}_k = \left[\frac{\partial \mathcal{L}_k}{\partial \mathbf{A}}, \frac{\partial \mathcal{L}_k}{\partial \mathbf{B}}, \frac{\partial \mathcal{L}_k}{\partial \mathbf{c}} \right]$. For this reason, the $\hat{\boldsymbol{\Theta}}_l$ updating equation requires the calculation of the likelihood function partial derivatives in respect to the model parameters. This computation is not a trivial task since they do not possess closed form and can only be obtained iteratively. Thus, in the following three parts, all the necessary calculations to derive those expressions are presented.

3.5.1.1 Partial derivative of $\mathcal{L}_k$ in order to a_{ij}

The updating of matrix $\mathbf{A}$ coefficients along the likelihood gradient direction, is obtained by iteratively executing the following expression:

$$\hat{\mathbf{A}}_{l+1} = \hat{\mathbf{A}}_l + \eta_a \left. \frac{\partial \mathcal{L}_k}{\partial \mathbf{A}} \right|_{\mathbf{A} = \hat{\mathbf{A}}_l} \qquad (3.145)$$

where $\eta_a > 0$ refers to the learning rate. The dimension of $\mathbf{A}$ is $m \times m$ which implies that $\frac{\partial \mathcal{L}_k}{\partial \mathbf{A}}$ must necessarily have the same size. In particular, it will have the following shape:

$$\frac{\partial \mathcal{L}_k}{\partial \mathbf{A}} = \begin{bmatrix} \frac{\partial \mathcal{L}_k}{\partial a_{11}} & \cdots & \frac{\partial \mathcal{L}_k}{\partial a_{1m}} \\ \vdots & \ddots & \vdots \\ \frac{\partial \mathcal{L}_k}{\partial a_{m1}} & \cdots & \frac{\partial \mathcal{L}_k}{\partial a_{mm}} \end{bmatrix} \tag{3.146}$$

For notation simplicity, the following variable attribution is introduced,

$$\Omega_a \left(\xi, \varepsilon \right) = \frac{\partial \mathcal{L}_k}{\partial a_{\xi\varepsilon}} \tag{3.147}$$

The question is how to obtain $\Omega \left(\xi, \varepsilon \right)$ for $1 \leq \xi \leq m$ and $1 \leq \varepsilon \leq m$. According to the definition of the hidden Markov model likelihood function,

$$\mathcal{L}_k = \sum_{i=1}^{m} \alpha_k \left(i \right) \tag{3.148}$$

where, as previously stated,

$$\alpha_k \left(i \right) = \begin{cases} b_i \left(\lambda_1 \right) \cdot c_i, & k = 1 \\ b_i \left(\lambda_k \right) \sum_{j=1}^{m} a_{ji} \alpha_{k-1} \left(j \right), & k \neq 1 \end{cases} \tag{3.149}$$

For an arbitrary parameter $a_{\xi\varepsilon}$ belonging to $\mathbf{A}$, then:

$$\frac{\partial \mathcal{L}_k}{\partial a_{\xi\varepsilon}} = \sum_{i=1}^{m} \frac{\partial \alpha_k(i)}{\partial a_{\xi\varepsilon}} \tag{3.150}$$

According to (3.149), then:

$$\frac{\partial \alpha_k \left(i \right)}{\partial a_{\xi\varepsilon}} = \begin{cases} 0, & k = 1 \\ b_i \left(\lambda_k \right) \sum_{j=1}^{m} \left(\frac{\partial a_{ji}}{\partial a_{\xi\varepsilon}} \alpha_{k-1} \left(j \right) + a_{ji} \frac{\partial \alpha_{k-1}(j)}{\partial a_{\xi\varepsilon}} \right), & k \neq 1 \end{cases} \tag{3.151}$$

If $i \neq \varepsilon$ then $\frac{\partial a_{ji}}{\partial a_{\xi\varepsilon}} = 0$ and thus,

$$\frac{\partial \alpha_k \left(i \right)}{\partial a_{\xi\varepsilon}} = b_i \left(\lambda_k \right) \sum_{j=1}^{m} a_{ji} \frac{\partial \alpha_{k-1} \left(j \right)}{\partial a_{\xi\varepsilon}}, \ k \neq 1 \tag{3.152}$$

If, on the other hand, $i = \varepsilon$, then $\frac{\partial a_{ji}}{\partial a_{\xi\varepsilon}}$ is only nonzero if $j = \xi$. Therefore,

$$\frac{\partial \alpha_k \left(i \right)}{\partial a_{\xi\delta}} = b_i \left(\lambda_k \right) \left(\alpha_{k-1} \left(\xi \right) + \sum_{j=1}^{m} a_{ji} \frac{\partial \alpha_{k-1} \left(j \right)}{\partial a_{\xi\delta}} \right), \ k \neq 1 \tag{3.153}$$

As can be seen, there is feedback in the above expression. That is, the forward probabilities derivatives are calculated recursively. More clearly, by making the following assignment,

$$\frac{\partial \alpha_k (i)}{\partial a_{\xi\varepsilon}} = \Phi_k (\xi, \varepsilon, i) \tag{3.154}$$

leads to:

$$\Phi_k (\xi, \varepsilon, i) = \begin{cases} 0 & k = 1 \\ b_i (\lambda_k) \sum_{j=1}^{m} a_{ji}\Phi_{k-1} (\xi, \varepsilon, j) & k \neq 1 \wedge i \neq \varepsilon \\ b_\varepsilon (\lambda_k) \left(\alpha_{k-1} (\xi) + \sum_{j=1}^{m} a_{j\varepsilon}\Phi_{k-1} (\xi, \varepsilon, j) \right) & k \neq 1 \wedge i = \varepsilon \end{cases} \tag{3.155}$$

and,

$$\Omega_a (\xi, \varepsilon) = \sum_{i=1}^{m} \Phi_k (\xi, \varepsilon, i) \tag{3.156}$$

This $\Omega_a(\xi, \varepsilon)$ computation strategy is summarized in the pseudo-code described at Algorithm 2.

Using the above mathematical relations, the partial derivative $\frac{\partial \mathcal{L}}{\partial \mathbf{A}}$ can then be computed in MATLAB® by means of the function presented in Listing 3.14. However, in practice, this formulation should be avoided since it will easily reach and exceed the computer's numerical resolution. Instead, a normalized version should be used. This phenomena is aligned with the problem of numeric instability mentioned during Section 3.3.2.3.

One way to circumvent this problem is to change the scale of α_k (and eventually β_k) at each iteration k. In other words, at each time instant k, the value of $\bar{\alpha}_k$ is obtained as a scaled version of α_k. The relationship between both variables is given by:

$$\bar{\alpha}_k (i) = \alpha_k (i) \prod_{j=1}^{k} u_j \tag{3.157}$$

where the coefficients u_j have already been defined during Section 3.3.2.3.

Recalling (3.150), and after taking into consideration the above expression, the derivative of the likelihood function over $a_{\xi\varepsilon}$, is:

$$\frac{\partial \mathcal{L}_k}{\partial a_{\xi\varepsilon}} = \frac{1}{\prod_{l=1}^{k} u_l} \sum_{i=1}^{m} \frac{\partial \bar{\alpha}_k (i)}{\partial a_{\xi\varepsilon}} = \frac{1}{\prod_{l=1}^{k} c_l} \sum_{i=1}^{m} \bar{\bar{\Phi}}_k (\xi, \varepsilon, i) \tag{3.158}$$

Listing 3.14 MATLAB® function to compute $\frac{\partial \mathcal{L}}{\partial \mathbf{A}}$.

```matlab
 1  function [ALFA,OMEGA]=gradientLA(A,B,O,c)
 2  % Likelihood gradient with respect to A
 3  % m hidden states, n output states and N observations
 4  % A - mxm (state transitions matrix)
 5  % B - nxm (confusion matrix)
 6  % O - 1xN (observations vector)
 7  % c - 1xm (priors vector)
 8
 9  [m,n]=size(B);              % n - Number of visible states
10  N=length(O);               % N - Dimension of the observed sequence
11  PHI=zeros(m,m,m);
12  ALFA=zeros(N,m);
13  OMEGA=zeros(m,m);
14  PHI2=zeros(m,m,m);
15  k=1;
16  for i=1:m,
17    ALFA(k,i)=B(i,O(k))*c(i);
18    for csi=1:m,
19      for epsilon=1:m,
20        PHI(csi,epsilon,i)=0;
21        OMEGA(csi,epsilon)=OMEGA(csi,epsilon)+PHI(csi,epsilon,i);
22      end
23    end
24  end
25  for k=2:N,
26    for i=1:m,
27      S=0;
28      for j=1:m,
29        S=S+A(j,i)*ALFA(k-1,j);
30      end
31      ALFA(k,i)=B(i,O(k))*S;
32      for csi=1:m,
33        for epsilon=1:m,
34          if i~=epsilon,
35            S=0;
36            for j=1:m,
37              S=S+A(j,i)*PHI(csi,epsilon,j);
38            end
39            PHI2(csi,epsilon,i)=B(i,O(k))*S;
40          else
41            S=0;
42            for j=1:m,
43              S=S+A(j,i)*PHI(csi,epsilon,j);
44            end
45            PHI2(csi,epsilon,i)=B(i,O(k))*(ALFA(k-1,csi)+S);
46          end
47          OMEGA(csi,epsilon)=OMEGA(csi,epsilon)+PHI2(csi,epsilon,i);
48        end
49      end
50    end
51    PHI=PHI2;
52  end
```

Algorithm 2 Algorithm for the calculation of $\Omega_a\left(\xi,\varepsilon\right)$

For $k = 1$ to N determine,
  If $k = 1$ then,
    For $i = 1$ to m determine,
      $\alpha_1\left(i\right) = b_i\left(\lambda_1\right) \cdot c_i$
      For $\xi = 1$ to m determine,
        For $\varepsilon = 1$ to m determine,
          $\Phi_k\left(\xi,\varepsilon,i\right) = 0$
          $\Omega_a\left(\xi,\varepsilon\right) \leftarrow \Omega_a\left(\xi,\varepsilon\right) + \Phi_k\left(\xi,\varepsilon,i\right)$
        Repeat
      Repeat
    Repeat
  Else,
    For $i = 1$ to m determine,
      $\bar{\alpha}_k\left(i\right) = b_i\left(\lambda_k\right) \cdot \sum_{j=1}^{m} a_{ji}\bar{\alpha}_{k-1}\left(j\right)$
      For $\xi = 1$ to m determine,
        For $\varepsilon = 1$ to m determine,
          If $i \neq \delta$ then,
            $\Phi_k\left(\xi,\varepsilon,i\right) = b_i\left(\lambda_k\right)\sum_{j=1}^{m} a_{ji}\Phi_{k-1}\left(\xi,\varepsilon,j\right)$
          Else,
            $\Phi_k\left(\xi,\varepsilon,i\right) = b_i\left(\lambda_k\right)\left(\alpha_{k-1}\left(\xi\right) + \sum_{j=1}^{m} a_{ji}\Phi_{k-1}\left(\xi,\varepsilon,j\right)\right)$
          Continue,
          $\Omega_a\left(\xi,\varepsilon\right) \leftarrow \Omega_a\left(\xi,\varepsilon\right) + \Phi_k\left(\xi,\varepsilon,i\right)$
        Repeat
      Repeat
    Repeat
  Continue,
Repeat

From an algebraic point-of-view, it is easier to compute the log-likelihood derivative. That is,

$$\frac{\partial}{\partial a_{\xi\varepsilon}}\ln\left(\mathcal{L}_k\right) = \frac{1}{\mathcal{L}_k \prod_{l=1}^{k} u_l}\sum_{i=1}^{m}\frac{\partial \bar{\alpha}_k\left(i\right)}{\partial a_{\xi\varepsilon}} \tag{3.159}$$

Notice that, since the logarithm is an increasing monotonous function, the value of $a_{\xi\varepsilon}$ that maximizes $\mathcal{L}_k$ is the same that maximizes $\ln\left(\mathcal{L}_k\right)$. On the other hand, since

$$\mathcal{L}_k = \frac{1}{\prod_{l=1}^{k} u_l} \tag{3.160}$$

then,

$$\frac{\partial}{\partial a_{\xi\varepsilon}}\ln\left(\mathcal{L}_k\right) = \bar{\Omega}\left(\xi,\varepsilon\right) = \sum_{i=1}^{m}\bar{\Phi}_k\left(\xi,\varepsilon,i\right) \tag{3.161}$$

As previously seen, the update expression of α_k is:

$$\alpha_k\left(i\right) = \begin{cases} b_i\left(\lambda_1\right) \cdot c_i, & k = 1 \\ b_i\left(\lambda_k\right)\sum_{j=1}^{m} a_{ji}\alpha_{k-1}\left(j\right), & k \neq 1 \end{cases} \tag{3.162}$$

Considering the case where $k \neq 1$, and after multiplying by $\prod_{l=1}^{k} u_l$ both terms of the previous equality, leads to:

$$\alpha_k(i) \cdot \prod_{l=1}^{k} u_l = b_i(\lambda_k) \cdot \prod_{l=1}^{k} u_l \cdot \sum_{j=1}^{m} a_{ji}\alpha_{k-1}(j) \qquad (3.163)$$

That is,

$$
\begin{aligned}
\alpha_k(i) \cdot \prod_{l=1}^{k} u_l &= b_i(\lambda_k) \cdot \sum_{j=1}^{m} a_{ji}\alpha_{k-1}(j) \cdot \prod_{l=1}^{k} u_l \\
&= b_i(\lambda_k) \cdot \sum_{j=1}^{m} a_{ji}\alpha_{k-1}(j) \cdot u_k \cdot \prod_{l=1}^{k-1} u_l \qquad (3.164) \\
&= b_i(\lambda_k) \cdot \sum_{j=1}^{m} a_{ji}\bar{\alpha}_{k-1}(j) \cdot u_k
\end{aligned}
$$

and finally,

$$\bar{\alpha}_k(i) = b_i(\lambda_k) \cdot u_k \cdot \sum_{j=1}^{m} a_{ji}\bar{\alpha}_{k-1}(j) \qquad (3.165)$$

Again, for an arbitrary coefficient $a_{\xi\varepsilon}$,

$$\frac{\partial}{\partial a_{\xi\varepsilon}}\bar{\alpha}_k(i) = b_i(\lambda_k) \cdot u_k \cdot \sum_{j=1}^{m} \frac{\partial}{\partial a_{\xi\varepsilon}}(a_{ji}\bar{\alpha}_{k-1}(j)) \qquad (3.166)$$

which, after applying the product rule for derivatives, leads to:

$$\frac{\partial}{\partial a_{\xi\varepsilon}}\bar{\alpha}_k(i) = b_i(\lambda_k) \cdot u_k \cdot \sum_{j=1}^{m} \left(\bar{\alpha}_{k-1}(j)\frac{\partial a_{ji}}{\partial a_{\xi\varepsilon}} + a_{ji}\frac{\partial \bar{\alpha}_{k-1}(j)}{\partial a_{\xi\varepsilon}}\right) \qquad (3.167)$$

Defining $\frac{\partial \bar{\alpha}_k(i)}{\partial a_{\xi\varepsilon}}$ as $\bar{\Phi}_k(\xi,\varepsilon,i)$ results in:

$$\bar{\Phi}_k(\xi,\varepsilon,i) = b_i(\lambda_k) \cdot u_k \cdot \sum_{j=1}^{m} \left(\bar{\alpha}_{k-1}(j)\frac{\partial a_{ji}}{\partial a_{\xi\varepsilon}} + a_{ji}\bar{\Phi}_{k-1}(\xi,\varepsilon,j)\right) \qquad (3.168)$$

which turns out to be:

$$\bar{\Phi}_k(\xi,\varepsilon,i) = \left\{ \begin{array}{l} b_i(\lambda_k) \cdot u_k \cdot \sum_{j=1}^{m} a_{ji}\bar{\Phi}_{k-1}(\xi,\varepsilon,j), \varepsilon \neq i \\ b_i(\lambda_k) \cdot u_k \cdot \left(\bar{\alpha}_{k-1}(\xi) + \sum_{j=1}^{m} a_{ji}\bar{\Phi}_{k-1}(\xi,\varepsilon,j)\right), \varepsilon = i \end{array} \right.$$
$$(3.169)$$

The equation for updating the transition probability from state ξ to state ε is thus,

$$[\hat{a}_{\xi\varepsilon}]_{l+1} = [\hat{a}_{\xi\varepsilon}]_l + \eta_a \left[\bar{\Omega}_a \left(\xi, \varepsilon\right)\right]_{a_{\xi\varepsilon} = [\hat{a}_{\xi\varepsilon}]_l} \qquad (3.170)$$

Where, as previously mentioned,

$$\bar{\Omega}_a \left(\xi, \varepsilon\right) = \sum_{i=1}^{m} \bar{\Phi}_k \left(\xi, \varepsilon, i\right) \qquad (3.171)$$

which, as can be seen from expression (3.169), depends only on the value of $\bar{\alpha}$, the scaled version of α. Algorithm 3 presents the sequence of operations to obtain $\bar{\Omega}_a \left(\xi, \epsilon\right)$.

Algorithm 3 Algorithm for the calculation of $\bar{\Omega}_a \left(\xi, \varepsilon\right)$

For $k = 1$ to N determine,
   If $k = 1$ then,
      For $i = 1$ to m determine,
         $\bar{\alpha}_1 \left(i\right) = u_1 \cdot b_i \left(\lambda_1\right) \cdot c_i$
            For $\xi = 1$ to m determine,
               For $\delta = 1$ to m determine,
                  $\bar{\Phi}_k \left(\xi, \varepsilon, i\right) = 0$
                  $\bar{\Omega}_a \left(\xi, \varepsilon\right) \leftarrow \bar{\Omega}_a \left(\xi, \varepsilon\right) + \bar{\Phi}_k \left(\xi, \varepsilon, i\right)$
               Repeat
            Repeat
      Repeat
   Else,
      For $i = 1$ to m determine,
         $\bar{\alpha}_k \left(i\right) = b_i \left(\lambda_k\right) \cdot u_k \cdot \sum_{j=1}^{m} a_{ji} \bar{\alpha}_{k-1} \left(j\right)$
         For $\xi = 1$ to m determine,
            For $\delta = 1$ to m determine,
               If $i \neq \delta$ then,
                  $\bar{\Phi}_k \left(\xi, \varepsilon, i\right) = b_i \left(\lambda_k\right) \cdot u_k \cdot \sum_{j=1}^{m} a_{ji} \bar{\Phi}_{k-1} \left(\xi, \varepsilon, j\right)$
               Else,
                  $\bar{\Phi}_k \left(\xi, \varepsilon, i\right) = b_i \left(\lambda_k\right) \cdot u_k \cdot \left(\bar{\alpha}_{k-1} \left(\xi\right) + \sum_{j=1}^{m} a_{ji} \bar{\Phi}_{k-1} \left(\xi, \varepsilon, j\right)\right)$
               Continue,
               $\bar{\Omega}_a \left(\xi, \varepsilon\right) \leftarrow \bar{\Omega}_a \left(\xi, \varepsilon\right) + \bar{\Phi}_k \left(\xi, \varepsilon, i\right)$
            Repeat
          Repeat
      Repeat
   Continue,
Repeat

The normalized version of the method to compute $\frac{\partial \mathcal{L}}{\partial \mathbf{A}}$, coded as a MATLAB® function, can be found in Listing 3.15.

Listing 3.15 Compute $\frac{\partial \mathcal{L}}{\partial \mathbf{A}}$ with normalization.

```
1  function [ALFA,OMEGA]=gradientLA_norm(A,B,O,c)
2  % Likelihood gradient with respect to A (normalized version)
3  % m hidden states, n output states and N observations
4  % A - mxm (state transitions matrix)
5  % B - nxm (confusion matrix)
6  % O - 1xN (observations vector)
7  % c - 1xm (priors vector)
8  % ALFA - Matrix of forward probabilities
9  % OMEGA - Matrix of the derivative of L in order to A
10
11 [m,~]=size(B);      % n - Number of visible states
12 N=length(O);        % N - Dimension of the observed sequence
13 PHI=zeros(m,m,m);
14 [ALFA,~,n]=forward_algorithm_norm(A,B,O,c);
15 OMEGA=zeros(m,m);
16 PHI2=zeros(m,m,m);
17 k=1;
18
19 for i=1:m,
20     ALFA(k,i)=n(k)*B(i,O(k))*c(i);
21     for csi=1:m,
22         for delta=1:m,
23             PHI(csi,delta,i)=0;
24             OMEGA(csi,delta)=OMEGA(csi,delta)+PHI(csi,delta,i);
25         end
26     end
27 end
28
29 for k=2:N,
30     for i=1:m,
31         S=0;
32         for j=1:m,
33             S=S+A(j,i)*ALFA(k-1,j);
34         end
35         ALFA(k,i)=n(k)*B(i,O(k))*S;
36         for csi=1:m,
37             for delta=1:m,
38                 if i~=delta,
39                     S=0;
40                     for j=1:m,
41                         S=S+A(j,i)*PHI(csi,delta,j);
42                     end
43                     PHI2(csi,delta,i)=B(i,O(k))*n(k)*S;
44                 else
45                     S=0;
46                     for j=1:m,
47                         S=S+A(j,i)*PHI(csi,delta,j);
48                     end
49                     PHI2(csi,delta,i)=B(i,O(k))*n(k)*(ALFA(k-1,csi)+S);
50                 end
51                 OMEGA(csi,delta)=OMEGA(csi,delta)+PHI2(csi,delta,i);
52             end
53         end
54     end
55     PHI=PHI2;
56 end
```

3.5.1.2 Partial derivative of $\mathcal{L}_k$ in order to b_{ij}

During this section, the same sequence of operations will be presented as in the previous section. However, at this time, the focus is on the parameter matrix **B**. The recursive formula for updating this matrix is given by:

$$\hat{\mathbf{B}}_{l+1} = \hat{\mathbf{B}}_l + \eta_b \left. \frac{\partial \mathcal{L}_k}{\partial \mathbf{B}} \right|_{\mathbf{B}=\hat{\mathbf{B}}_l} \tag{3.172}$$

Since the dimension of **B** is $m \times n$, then $\frac{\partial \mathcal{L}_k}{\partial \mathbf{B}}$ must have the following structure:

$$\frac{\partial \mathcal{L}_k}{\partial \mathbf{B}} = \begin{bmatrix} \frac{\partial \mathcal{L}_k}{\partial b_{11}} & \cdots & \frac{\partial \mathcal{L}_k}{\partial b_{1n}} \\ \vdots & \ddots & \vdots \\ \frac{\partial \mathcal{L}_k}{\partial b_{m1}} & \cdots & \frac{\partial \mathcal{L}_k}{\partial b_{mn}} \end{bmatrix} \tag{3.173}$$

As before, in order to simplify the notation, the following equality is considered:

$$\Omega_b\left(\xi, \varepsilon\right) = \frac{\partial \mathcal{L}_k}{\partial b_{\xi\varepsilon}} \tag{3.174}$$

Considering (3.148), and after derivation in respect to $b_{\xi\varepsilon}$, the following expression is obtained:

$$\frac{\partial \mathcal{L}_k}{\partial b_{\xi\varepsilon}} = \sum_{i=1}^{m} \frac{\partial \alpha_k(i)}{\partial b_{\xi\varepsilon}} \tag{3.175}$$

Now, for $k = 1$ and assuming that $\lambda_1 = r_\varepsilon$,

$$\frac{\partial \alpha_k(i)}{\partial b_{\xi\varepsilon}} = \frac{\partial}{\partial b_{\xi\varepsilon}}\left(b_{i\varepsilon} \cdot c_i\right) = \begin{cases} 0, & i \neq \xi \\ c_i, & i = \xi \end{cases} \tag{3.176}$$

On the other hand, for $k = 1$, and considering that $\lambda_1 \neq r_\varepsilon$,

$$\frac{\partial \alpha_k(i)}{\partial b_{\xi\varepsilon}} = 0 \tag{3.177}$$

Which, for $k > 1$,

$$\frac{\partial \alpha_k(i)}{\partial b_{\xi\varepsilon}} = \frac{\partial}{\partial b_{\xi\varepsilon}}\left(b_i\left(\lambda_k\right) \sum_{j=1}^{m} a_{ji}\alpha_{k-1}(j)\right)$$

$$= \frac{\partial b_i\left(\lambda_k\right)}{\partial b_{\xi\varepsilon}} \sum_{j=1}^{m} a_{ji}\alpha_{k-1}(j) + b_i\left(\lambda_k\right) \frac{\partial}{\partial b_{\xi\varepsilon}} \sum_{j=1}^{m} a_{ji}\alpha_{k-1}(j) \tag{3.178}$$

$$= \frac{\partial b_i\left(\lambda_k\right)}{\partial b_{\xi\varepsilon}} \sum_{j=1}^{m} a_{ji}\alpha_{k-1}(j) + b_i\left(\lambda_k\right) \sum_{j=1}^{m} a_{ji}\frac{\partial \alpha_{k-1}(j)}{\partial b_{\xi\varepsilon}}$$

If $\lambda_k = r_\varepsilon$,

$$\frac{\partial \alpha_k (i)}{\partial b_{\xi\varepsilon}} = \frac{\partial b_{i\delta}}{\partial b_{\xi\varepsilon}} \sum_{j=1}^{m} a_{ji} \alpha_{k-1} (j) + b_{i\varepsilon} \sum_{j=1}^{m} a_{ji} \frac{\partial \alpha_{k-1} (j)}{\partial b_{\xi\varepsilon}} \tag{3.179}$$

Still within the condition of $\lambda_k = r_\varepsilon$, if $i = \xi$

$$\frac{\partial \alpha_k (i)}{\partial b_{\xi\varepsilon}} = \sum_{j=1}^{m} a_{j\xi} \alpha_{k-1} (j) + b_{\xi\varepsilon} \sum_{j=1}^{m} a_{j\xi} \frac{\partial \alpha_{k-1} (j)}{\partial b_{\xi\varepsilon}} \tag{3.180}$$

On the other hand, if $i \neq \xi$

$$\frac{\partial \alpha_k (i)}{\partial b_{\xi\varepsilon}} = b_{i\varepsilon} \sum_{j=1}^{m} a_{ji} \frac{\partial \alpha_{k-1} (j)}{\partial b_{\xi\varepsilon}} \tag{3.181}$$

If now $\lambda_k \neq r_\delta$ then,

$$\frac{\partial \alpha_k (i)}{\partial b_{\xi\varepsilon}} = b_i (\lambda_k) \sum_{j=1}^{m} a_{ji} \frac{\partial \alpha_{k-1} (j)}{\partial b_{\xi\varepsilon}} \tag{3.182}$$

Finally, performing the assignment:

$$\frac{\partial \alpha_k (i)}{\partial b_{\xi\varepsilon}} = \Psi_k (\xi, \varepsilon, i) \tag{3.183}$$

and considering that,

$$\Omega_b (\xi, \delta) = \sum_{i=1}^{m} \Psi_k (\xi, \delta, i) \tag{3.184}$$

leads to Algorithm 4 which describes the pseudo-code that can be used to compute the above expression.

Finally, having $\Omega_b(\xi, \varepsilon)$, the gradient of $\frac{\partial \mathcal{L}}{\partial \mathbf{B}}$ can be computed, in a straightforward manner, as shown in the MATLAB® function presented at Listing 3.16.

It is worth to remember that, in practice and due to numeric stability issues, the forward probability calculation is performed by means of parameter scaling. Hence, the mathematical conclusions obtained earlier must be reformulated in order to obtain a more numerically robust expressions. In particular, the previous expressions must be expressed only as a function of $\bar{\alpha}_k$.

Listing 3.16 Compute $\frac{\partial \mathcal{L}}{\partial \mathbf{B}}$.

```
1  function [ALFA,OMEGA]=gradientLB(A,B,O,c)
2  % Likelihood gradient with respect to B
3  % m hidden states, n output states and N observations
4  % A - mxm (state transitions matrix)
5  % B - nxm (confusion matrix)
6  % O - 1xN (observations vector)
7  % c - 1xm (priors vector)
8  % ALFA - Matrix of direct probabilities
9  % OMEGA - Matrix of the derivative of L in order to B
10
11 [m,n]=size(B);              % n - Number of visible states
12 N=length(O);               % N - Dimension of the observed sequence
13 PHI=zeros(m,n,m);
14 ALFA=zeros(N,m);
15 OMEGA=zeros(m,n);
16 PHI2=zeros(m,n,m);
17 k=1;
18
19 for i=1:m,
20     ALFA(k,i)=B(i,O(k))*c(i);
21     for csi=1:m,
22         for delta=1:n,
23             if O(k)==delta,
24                 if i==csi,
25                     PHI(csi,delta,i)=c(i);
26                 else
27                     PHI(csi,delta,i)=0;
28                 end
29             else
30                 PHI(csi,delta,i)=0;
31             end
32             OMEGA(csi,delta)=OMEGA(csi,delta)+PHI(csi,delta,i);
33         end
34     end
35 end
36
37 for k=2:N,
38     for i=1:m,
39         %---------------------------Computes ALFA----------
40         S=0;
41         for j=1:m,
42             S=S+A(j,i)*ALFA(k-1,j);
43         end
44         %-----------------------------------------------------
45         ALFA(k,i)=B(i,O(k))*S;
46         for csi=1:m,
47             for delta=1:n,
48                 if O(k)==delta,
49                     if i==csi,
50                         S1=0;
51                         S2=0;
52                         for j=1:m,
53                             S1=S1+A(j,csi)*ALFA(k-1,j);
54                             S2=S2+A(j,csi)*PHI(csi,delta,j);
55                         end
56                         PHI2(csi,delta,i)=S1+B(csi,O(delta))*S2;
57                     else
```

```
58          S2=0;
59          for  j=1:m,
60              S2=S2+A(j,i)*PHI(csi,delta,j);
61          end
62          PHI2(csi,delta,i)=B(i,O(delta))*S2;
63        end
64      else
65          S2=0;
66          for  j=1:m,
67              S2=S2+A(j,i)*PHI(csi,delta,j);
68          end
69          PHI2(csi,delta,i)=B(i,O(k))*S2;
70        end
71        OMEGA(csi,delta)=OMEGA(csi,delta)+PHI2(csi,delta,i);
72      end
73    end
74  end
75  PHI=PHI2;
76 end
```

Algorithm 4 Algorithm for the calculation of $\Omega_b\left(\xi, \varepsilon\right)$

For $k = 1$ to N determine,
    If $k = 1$ then,
        For $i = 1$ to m determine,
            For $\xi = 1$ to m,
                For $\varepsilon = 1$ to n,
                    If $\lambda_1 = r_\varepsilon$ then,
                        If $i = \xi$ then,
                            $\Psi_k\left(\xi, \varepsilon, i\right) = c_i$
                        Else,
                            $\Psi_k\left(\xi, \varepsilon, i\right) = 0$
                          Continue,
                    Else,
                      $\Psi_k\left(\xi, \varepsilon, i\right) = 0$
                    Continue,
                    $\Omega_b\left(\xi, \varepsilon\right) \leftarrow \Omega_b\left(\xi, \varepsilon\right) + \Psi_k\left(\xi, \varepsilon, i\right)$
                Repeat
            Repeat
        Repeat
    Else,
        For $i = 1$ to m determine,
            For $\xi = 1$ to m,
                 For $\varepsilon = 1$ to n,
                    If $\lambda_k = r_\varepsilon$ then,
                      If $i = \xi$ then,
                          $\Psi_k\left(\xi, \varepsilon, i\right) = \sum_{j=1}^{m} a_{j\xi}\alpha_{k-1}\left(j\right) +$
                        $b_{\xi\delta}\sum_{j=1}^{m} a_{j\xi}\Psi_{k-1}\left(\xi, \varepsilon, j\right)$
                      Else,
                          $\Psi_k\left(\xi, \varepsilon, i\right) = b_{i\varepsilon}\sum_{j=1}^{m} a_{ji}\Psi_{k-1}\left(\xi, \varepsilon, j\right)$
                        Continue,
                    Else,
                      $\Psi_k\left(\xi, \varepsilon, i\right) = b_i\left(\lambda_k\right)\sum_{j=1}^{m} a_{ji}\Psi\left(\xi, \varepsilon, j\right)$
                    Continue,
                    $\Omega_b\left(\xi, \varepsilon\right) \leftarrow \Omega_b\left(\xi, \varepsilon\right) + \Psi_k\left(\xi, \varepsilon, i\right)$
                Repeat
            Repeat
        Repeat
    Continue
Repeat

Notice that the likelihood function can be obtained from the forward probabilities by:

$$\mathcal{L}_k = \sum_{i=1}^{m} \alpha_k(i) \tag{3.185}$$

From the re-scaling procedure described in Section 3.3.2.3, the forward probability is related to its scaled version by:

$$\bar{\alpha}_k(i) = \alpha_k(i) \prod_{l=1}^{k} n_l \tag{3.186}$$

leading to:

$$\mathcal{L}_k = \frac{1}{\prod_{l=1}^{k} n_l} \sum_{i=1}^{m} \bar{\alpha}_k(i) \tag{3.187}$$

Taking into consideration the above equalities, the derivative of the likelihood function, in respect of $b_{\xi\varepsilon}$, is given by:

$$\frac{\partial \mathcal{L}_k}{\partial b_{\xi\varepsilon}} = \frac{1}{\prod_{l=1}^{k} n_l} \sum_{i=1}^{m} \frac{\partial \bar{\alpha}_k(i)}{\partial b_{\xi\varepsilon}} = \mathcal{L}_k \sum_{i=1}^{m} \frac{\partial \bar{\alpha}_k(i)}{\partial b_{\xi\varepsilon}} \tag{3.188}$$

Usually, due to numerical stability issues, it is not possible to compute $\mathcal{L}_k$, and so the derivative of the log-likelihood is used instead. This replacement can be done since the set of parameters that improves $\mathcal{L}_k$ also improves $\ln(\mathcal{L}_k)$. For this reason, the gradient of the logarithm of $\mathcal{L}_k$ along $b_{\xi\varepsilon}$ is defined as:

$$\frac{\partial}{\partial b_{\xi\varepsilon}} \ln(\mathcal{L}_k) = \sum_{i=1}^{m} \frac{\partial \bar{\alpha}_k(i)}{\partial b_{\xi\varepsilon}} \tag{3.189}$$

For $k > 1$,

$$\bar{\alpha}_k(i) = b_i(\lambda_k) \cdot n_k \cdot \sum_{j=1}^{m} a_{ji} \bar{\alpha}_{k-1}(j) \tag{3.190}$$

The partial derivative of $\bar{\alpha}_k(i)$ in order to $b_{\xi\varepsilon}$ is then,

$$\frac{\partial \bar{\alpha}_k(i)}{\partial b_{\xi\varepsilon}} =$$
$$n_k \frac{\partial b_i(\lambda_k)}{\partial b_{\xi\varepsilon}} \cdot \sum_{j=1}^{m} a_{ji} \bar{\alpha}_{k-1}(j) + n_k b_i(\lambda_k) \cdot \sum_{j=1}^{m} a_{ji} \frac{\partial \bar{\alpha}_{k-1}(j)}{\partial b_{\xi\varepsilon}} \tag{3.191}$$

Defining $\bar{\Psi}_k(\xi, \varepsilon, i)$ to be equal to:

$$\bar{\Psi}_k(\xi, \varepsilon, i) = \frac{\partial \bar{\alpha}_k(i)}{\partial b_{\xi\varepsilon}} \tag{3.192}$$

leads to:

$$\bar{\Psi}_k(\xi, \varepsilon, i) =$$
$$\frac{\partial b_i(\lambda_k)}{\partial b_{\xi\varepsilon}} \cdot n_k \cdot \sum_{j=1}^{m} a_{ji}\bar{\alpha}_{k-1}(j) + b_i(\lambda_k) \cdot n_k \cdot \sum_{j=1}^{m} a_{ji}\bar{\Psi}_{k-1}(\xi, \varepsilon, j) \tag{3.193}$$

When $i \neq \xi$

$$\bar{\Psi}_k(\xi, \varepsilon, i) = b_i(\lambda_k) \cdot n_k \cdot \sum_{j=1}^{m} a_{ji}\bar{\Psi}_{k-1}(\xi, \varepsilon, j) \tag{3.194}$$

On the other hand, when $i = \xi$, there are two situations to evaluate: if $\lambda_k = r_\varepsilon$ or if $\lambda_k \neq r_\varepsilon$. Regarding the first, $\bar{\Psi}_k(\xi, \varepsilon, i)$ is equal to:

$$\bar{\Psi}_k(\xi, \varepsilon, i) = c_k \cdot \sum_{j=1}^{m} a_{j\xi}\bar{\alpha}_{k-1}(j) + b_{\xi\varepsilon} \cdot n_k \cdot \sum_{j=1}^{m} a_{j\xi}\bar{\Psi}_{k-1}(\xi, \varepsilon, j) \tag{3.195}$$

and for the second:

$$\bar{\Psi}_k(\xi, \varepsilon, i) = b_\xi(\lambda_k) \cdot n_k \cdot \sum_{j=1}^{m} a_{j\xi}\bar{\Psi}_{k-1}(\xi, \varepsilon, j) \tag{3.196}$$

Before concluding it is now necessary to analyze the initial situation. It is known that, for $k = 1$,

$$\bar{\alpha}_1(i) = n_1 \cdot b_i(\lambda_1) \cdot c_i \tag{3.197}$$

resulting in,

$$\bar{\Psi}_1(\xi, \varepsilon, i) = n_1 \cdot \frac{\partial b_i(\lambda_1)}{\partial b_{\xi\varepsilon}} \cdot c_i \tag{3.198}$$

If $i = \xi$ and $\lambda_1 = r_\varepsilon$ then,

$$\bar{\Psi}_1(\xi, \varepsilon, i) = n_1 \cdot c_i \tag{3.199}$$

For all other situations,

$$\bar{\Psi}_1(\xi, \varepsilon, i) = 0 \tag{3.200}$$

Algorithm 5 presents the pseudo-code that enables the computation of $\bar{\Omega}_b(\xi, \varepsilon)$. Moreover, in Listing 3.17, a MATLAB® function is presented that can be used to compute $\frac{\partial \bar{\mathcal{L}}_k}{b_{\xi\varepsilon}}$.

Listing 3.17 Computes $\frac{\partial \mathcal{L}}{\partial \mathbf{B}}$ with normalization.

```
1  function [ALFA,OMEGA]=gradientLB_norm(A,B,O,c)
2  % Likelihood gradient with respect to B (normalized version)
3  % m hidden states, n output states and N observations
4  % A - mxm (state transitions matrix)
5  % B - nxm (confusion matrix)
6  % O - 1xN (observations vector)
7  % c - 1xm (priors vector)
8  % ALFA - Matrix of forward probabilities
9  % OMEGA - Matrix of the derivative of L in order to B
10
11 [m,l]=size(B);       % n - Number of visible states
12 N=length(O);         % N - Dimension of the observed sequence
13 PHI=zeros(m,l,m);
14 [ALFA,~,n]=forward_algorithm_norm(A,B,O,c);
15 OMEGA=zeros(m,l);
16 PHI2=zeros(m,l,m);
17 k=1;
18
19 for i=1:m,
20     for csi=1:m,
21         for delta=1:l,
22             if O(k)==delta,
23                 if i==csi,
24                     PHI(csi,delta,i)=n(1)*c(i);
25                 else
26                     PHI(csi,delta,i)=0;
27                 end
28             else
29                 PHI(csi,delta,i)=0;
30             end
31             OMEGA(csi,delta)=OMEGA(csi,delta)+PHI(csi,delta,i);
32         end
33     end
34 end
35
36 for k=2:N,
37     for i=1:m,
38         for csi=1:m,
39             for delta=1:l,
40                 if i==csi,
41                     if O(k)==delta,
42                         S1=0;
43                         S2=0;
44                         for j=1:m,
45                             S1=S1+A(j,csi)*ALFA(k-1,j);
46                             S2=S2+A(j,csi)*PHI(csi,delta,j);
47                         end
48                         PHI2(csi,delta,i)=n(k)*S1+B(csi,delta)*n(k)*S2
                            ;
49                     else
50                         S2=0;
51                         for j=1:m,
52                             S2=S2+A(j,csi)*PHI(csi,delta,j);
53                         end
54                         PHI2(csi,delta,i)=B(csi,O(k))*n(k)*S2;
55                     end
56                 else
57                     S2=0;
58                     for j=1:m,
59                         S2=S2+A(j,i)*PHI(csi,delta,j);
60                     end
```

```
61                      PHI2(csi,delta,i)=B(i,O(k))*n(k)*S2;
62                  end
63                  OMEGA(csi,delta)=OMEGA(csi,delta)+PHI2(csi,delta,i);
64              end
65          end
66      end
67      PHI=PHI2;
68  end
```

Algorithm 5 Algorithm for the calculation of $\bar{\Omega}_b\left(\xi,\varepsilon\right)$

For $k = 1$ to N determine,
    If $k = 1$ then,
        For $i = 1$ to m determine,
            For $\xi = 1$ to m,
                For $\varepsilon = 1$ to n,
                    If $\lambda_1 = r_\varepsilon$ then,
                        If $i = \xi$ then,
                            $\bar{\Psi}_k\left(\xi,\varepsilon,i\right) = n_1 \cdot c_i$
                        Else,
                            $\bar{\Psi}_k\left(\xi,\varepsilon,i\right) = 0$
                        Continue,
                    Else,
                        $\bar{\Psi}_k\left(\xi,\varepsilon,i\right) = 0$
                    Continue,
                    $\bar{\Omega}_b\left(\xi,\varepsilon\right) \leftarrow \bar{\Omega}_b\left(\xi,\varepsilon\right) + \bar{\Psi}_k\left(\xi,\varepsilon,i\right)$
                Repeat
            Repeat
        Repeat
    Else,
        For $i = 1$ to m determine,
            For $\xi = 1$ to m,
                For $\delta = 1$ to n,
                    If $i = \xi$ then,
                        If $\lambda_k = r_\varepsilon$ then,
                            $\bar{\Psi}_k\left(\xi,\varepsilon,i\right) = n_k \cdot \sum_{j=1}^m a_{j\xi}\bar{\alpha}_{k-1}\left(j\right) + b_{\xi\varepsilon} \cdot n_k \cdot \sum_{j=1}^m a_{j\xi}\bar{\Psi}_{k-1}\left(\xi,\varepsilon,j\right)$
                        Else,
                            $\bar{\Psi}_k\left(\xi,\varepsilon,i\right) = b_\xi\left(\lambda_k\right) \cdot n_k \cdot \sum_{j=1}^m a_{j\xi}\bar{\Psi}_{k-1}\left(\xi,\varepsilon,j\right)$
                        Continue,
                    Else,
                        $\bar{\Psi}_k\left(\xi,\varepsilon,i\right) = b_i\left(\lambda_k\right) \cdot n_k \cdot \sum_{j=1}^m a_{ji}\bar{\Psi}_{k-1}\left(\xi,\varepsilon,j\right)$
                    Continue,
                    $\bar{\Omega}_b\left(\xi,\delta\right) \leftarrow \bar{\Omega}_b\left(\xi,\varepsilon\right) + \bar{\Psi}_k\left(\xi,\varepsilon,i\right)$
                Repeat
            Repeat
        Repeat
    Continue
Repeat

3.5.2 PARTIAL DERIVATIVE OF $\mathcal{L}_K$ IN ORDER TO c

Finally it remains to consider the case of the initial Markov chain probability distribution. Let c_i be the initial activation probability associated to the hidden state i and $\mathbf{c}$ the vector whose components are all the c_i for $i = 1$ up to the number of hidden states in the chain.

97

The gradient based updating rule for vector $\mathbf{c}$ can be expressed as:

$$\hat{\mathbf{c}}_{l+1} = \hat{\mathbf{c}}_l + \left. \frac{\partial \mathcal{L}_k}{\partial \mathbf{c}} \right|_{\mathbf{c}=\hat{\mathbf{c}}_l} \tag{3.201}$$

where

$$\frac{\partial \mathcal{L}_k}{\partial \mathbf{c}} = \left[\frac{\partial \mathcal{L}_k}{\partial c_1} \quad \cdots \quad \frac{\partial \mathcal{L}_k}{\partial c_m} \right] \tag{3.202}$$

Given that,

$$\mathcal{L}_k = \sum_{i=1}^{m} \alpha_k(i) \tag{3.203}$$

the gradient of this expression, according to an arbitrary parameter c_ξ, is:

$$\frac{\partial \mathcal{L}_k}{\partial c_\xi} = \sum_{i=1}^{m} \frac{\partial \alpha_k(i)}{\partial c_\xi} \tag{3.204}$$

Since, from the universe of all possible values for k, only $\alpha_1(\xi)$ depends on c_i, then,

$$\frac{\partial \mathcal{L}_1}{\partial c_\xi} = \sum_{i=1}^{m} \frac{\partial \alpha_1(i)}{\partial c_\xi} \tag{3.205}$$

That is,

$$\begin{aligned} \frac{\partial \mathcal{L}_1}{\partial c_\xi} &= 0 + \cdots + \frac{\partial \alpha_1(\xi)}{\partial c_\xi} + \cdots + 0 \\ &= b_\xi(\lambda_1) \end{aligned} \tag{3.206}$$

This leads to the following equation for the re-estimation of c_ξ:

$$\begin{aligned} [\hat{c}_\xi]_{l+1} &= [\hat{c}_\xi]_l + \eta_c \left[\frac{\partial \mathcal{L}_1}{\partial c_\xi} \right]_{c_\xi=[\hat{c}_\xi]_l} \\ &= [\hat{c}_\xi]_l + \eta_c b_\xi(\lambda_1) \end{aligned} \tag{3.207}$$

Algorithm 6 presents the pseudo-code for computing $\frac{\partial \mathcal{L}_1}{\partial c_\xi}$ and Listing 3.18 the respective MATLAB® code.

Algorithm 6 Algorithm for the calculation of $\frac{\partial \mathcal{L}_k}{\partial \mathbf{c}}$.

$k = 1$
For $i = 1$ to m determine,
    For $\xi = 1$ to m determine,
        If $i = \xi$ then,
            $c(i) = b_\xi(\lambda_1)$
        Else,
            $c(i) = 0$
        Continue,
    Repeat,
Repeat

Listing 3.18 MATLAB® function to compute $\frac{\partial \mathcal{L}_k}{\partial \mathbf{c}}$.

```
1  function [c,dLdc]=gradientLc(A,B,O,c)
2  % Likelihood gradient with respect to c
3  % m hidden states, n output states and N observations
4  % A - mxm (state transitions matrix)
5  % B - nxm (confusion matrix)
6  % O - 1xN (observations vector)
7  % c - 1xm (priors vector)
8
9  [m,~]=size(B);          % n - Number of visible states
10 dLdc=zeros(1,m);
11 k=1;
12 for i=1:m,
13     for xi=1:m,
14         if i==xi,
15             dLdc(i)=B(xi,O(k));
16         else
17             dLdc(i)=0;
18         end
19     end
20 end
```

3.5.3 PERFORMANCE ANALYSIS OF THE RE-ESTIMATION FORMULAS

During this section, a test example will be presented aiming to evaluate the capability of the re-estimation formulas, introduced in previous sections, to gradually find better parameters. For this, a hidden Markov model is defined with two hidden and three observable states with the following transition probabilities:

$$\mathbf{c} = \begin{bmatrix} 0.6 & 0.4 \end{bmatrix}$$

$$\mathbf{A} = \begin{bmatrix} 0.7 & 0.3 \\ 0.4 & 0.6 \end{bmatrix}$$

$$\mathbf{B} = \begin{bmatrix} 0.1 & 0.4 & 0.5 \\ 0.7 & 0.2 & 0.1 \end{bmatrix}$$

From this model, 2 sets of 20 observations each are drawn. The MAT-LAB® code used to generate these data sets is presented in Listing 3.19. Notice that this function makes a call to Listing 3.20.

Using this data, and by resorting to the gradient equations derived in the previous sections, an estimate of each of the probability transition matrices is obtained. This is done by the MATLAB® function presented in Listing 3.21 and using the previously referred data set. The obtained results, concerning the evolution of the model likelihood as a function of the iteration number, are illustrated in Figure 3.15. In particular, it presents the likelihood evolution considering only independent updates in the $\mathbf{A}$, $\mathbf{B}$ and $\mathbf{c}$ data structures. Notice that a total of 100 iterations, using a learning rate η equal to 0.005, was considered and the initial estimates for the probability transition matrices were randomly generated using an uniform probability distribution.

Listing 3.19 MATLAB® code used to generate the parameters estimation data set.

```
 1  function O=generate_data()
 2
 3  % Function used to generate training data from the HMM
 4  % with parameters:
 5  c=[0.6;0.4];
 6  A=[0.7 0.3;0.4 0.6];
 7  B=[0.1 0.4 0.5;0.7 0.2 0.1];
 8
 9  % Data generation considering route_nbr different routes
10  % and observ_nbr observations per route
11  route_nbr=2;
12  observ_nbr=20;
13
14  % Initialize matrices:
15  H=zeros(route_nbr,observ_nbr);  % Hidden
16  O=zeros(route_nbr,observ_nbr);  % Observable
17
18  for route=1:route_nbr,
19      H(route,1)=random_func(c);
20      for p=2:observ_nbr,
21          H(route,p)=random_func(A(H(route,p-1),:));
22      end
23      for p=1:observ_nbr,
24          O(route,p)=random_func(B(H(route,p),:));
25      end
26  end
```

Listing 3.20 Arbitrary distribution random number generator.

```
 1  function roulette=random_func(prob,N)
 2  % N - Number of elements
 3  % prob - Probabilities distribution: vector 1xn with distinct
 4  %        probability distribution
```

```
 5 % roulette - Vector 1xN with values within 1 and n according
 6 %              to distribution prob
 7 if nargin==1,
 8     N=1;
 9 end
10 % roulette wheel where each slice is proportional to
11 % the probability
12 L=length(prob);
13 prob=round(100*prob);
14 % roulette - vector with 100 elements whose distribution
15 % depends on P. For example if P=[0.3 0.7] then:
16 % roulette = [1 1 ... 1 2 2 ...2...2]
17 %            \------/ \---------/
18 %              30          70
19 roulette=[];
20 for k=1:L,
21     roulette=[roulette k*ones(1,prob(k))];
22 end
23
24 % Generates N values evenly distributed between 1 and 100
25 % (it will be the index of the "roulette" vector)
26 ptr=round(99*rand(1,N)+1);
27 roulette=roulette(ptr);
```

Listing 3.21 Function used to obtain the evolution of the model likelihood as a function of the iteration number.

```
 1 function test_gradient(O)
 2
 3 % First considering only update in matrix A. B and c are known and
 4 % A is randomly initialized.
 5 A=rand(2,2);
 6 A=A./(sum(A,2)*ones(1,2));
 7 B=[0.1 0.4 0.5;0.7 0.2 0.1];
 8 c=[0.6;0.4];
 9
10 % Learning rate
11 eta=0.005;
12
13 % Initialize likelihood vector. Each row is related to one of the
14 % 3 probability matrices/vectors
15 Likelihood=zeros(3,100);
16
17 % Iterations
18 for j=1:100
19     [~,OMEGA]=gradientLA_norm(A,B,O,c); % Compute gradient
20                                         % (normalized version)
21     A=A+eta*OMEGA; % Update matrix A
22     [~,lik]=forward_algorithm_norm(A,B,O,c); % Compute performance
23     Likelihood(1,j)=lik;
24 end
25
26 % Second, considering only update in matrix B. A and c are known
27 % and B is randomly initialized.
28 A=[0.7 0.3;0.4 0.6];
29 B=rand(2,3);
30 B=B./(sum(B,2)*ones(1,3));
31 c=[0.6;0.4];
32
```

```
33  % Iterations
34  for j=1:100
35      [~,OMEGA]=gradientLB_norm(A,B,O,c); % Compute gradient
36                                          % (normalized version)
37      B=B+eta*OMEGA; % Update matrix B
38      [~,lik]=forward_algorithm_norm(A,B,O,c); % Compute performance
39      Likelihood(2,j)=lik;
40  end
41
42  % Third, considering only update in vector c. A and B are known
43  % and c is randomly initialized.
44  A=[0.7 0.3;0.4 0.6];
45  B=[0.1 0.4 0.5;0.7 0.2 0.1];
46  c=rand(2,1);
47  c=c./sum(c);
48
49  % Iterations
50  for j=1:100
51      [dLdc]=gradientLc(B,O); % Compute gradient
52      c=c+eta*dLdc; % Update vector c
53      [~,lik]=forward_algorithm_norm(A,B,O,c); % Compute performance
54      Likelihood(3,j)=lik;
55  end
56
57  % Final plots
58  plot(1:100,Likelihood(1,:));
59  xlabel('Iteration');
60  ylabel('Log-Likelihood');
61  grid on;
62  figure
63  plot(1:100,Likelihood(2,:));
64  xlabel('Iteration');
65  ylabel('Log-Likelihood');
66  grid on;
67  figure
68  plot(1:100,Likelihood(3,:));
69  xlabel('Iteration');
70  ylabel('Log-Likelihood');
71  grid on
```

As can be seen from Figure 3.16, the hidden Markov model gradient based updating rules lead to a systematic increase in the model likelihood. However, there is a problem that has not been addressed so far. The stochasticity constraint that the transition probability matrices must obey. It is important to underline that the gradient based algorithm does not consider any restrictions in the parameters. Hence, it is necessary to investigate the applicability of an alternative optimization method that takes into account the constraints imposed on the parameter space. One of the alternatives is a re-parameterization of the parameters in order to guarantee the stochasticity condition. Another hypothesis is to use an alternative gradient based algorithm that is able to include those stochasticity conditions. In particular, the Rosen algorithm, which will be ad-

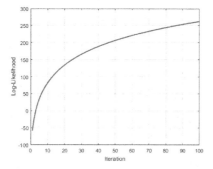

(a) Likelihood based on updating only the matrix **A**.

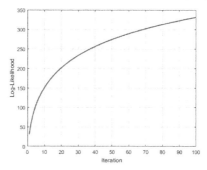

(b) Likelihood based on updating only the matrix **B**.

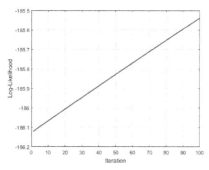

(c) Likelihood based on updating only the vector **c**.

Figure 3.16 Evolution of the performance of the Markov model, over 100 iterations, evaluated according to each of the parameters of the model independently.

dressed later in this chapter, has the ability to handle constraints into the parameters space. However, first the re-parameterization approach will be considered.

3.5.4 PARAMETERS COERCION BY RE-PARAMETERIZATION

The hidden Markov model parameters estimation method previously presented, does not consider the stochasticity constraint of the state probability transition matrices. In order to overcome this problem, it is possible to use an alternative parameter mapping by resorting to a monotonous function such as the sigmoidal function. For example, the state transition probability, from hidden state i to hidden state j, can be expressed by,

$$a_{ij} = \frac{e^{w_{ij}^a}}{\sum_{l=1}^{m} e^{w_{il}^a}} \tag{3.208}$$

the transition probability, from hidden state i to observable state j, as,

$$b_{ij} = \frac{e^{w_{ij}^b}}{\sum_{l=1}^{m} e^{w_{il}^b}} \tag{3.209}$$

and the priors by,

$$c_i = \frac{e^{w_i^c}}{\sum_{l=1}^{m} e^{w_l^c}} \tag{3.210}$$

where $e = 2.7182...$ refers to the Napier's constant and the domain of w_{ij}^a, w_{ij}^b and w_i^c is $\mathbb{R}$. It is straightforward to show that, with this parameterization,

$$\sum_{l=1}^{m} a_{il} = 1 \tag{3.211a}$$

$$\sum_{l=1}^{m} b_{il} = 1 \tag{3.211b}$$

$$\sum_{l=1}^{m} c_l = 1 \tag{3.211c}$$

which guarantees the stochastic conditions for the transition probabilities. Moreover, given that $e^{w_{ij}}$ is always positive and since $e^{w_{ij}} \leq \sum_{l=1}^{m} e^{w_{il}}$ then $0 < a_{ij} \leq 1$, $0 < b_{ij} \leq 1$ and $0 < c_i \leq 1$.

The gradient based learning law is:

$$[w_{ij}]_{l+1} = [w_{ij}]_l + \eta \cdot \frac{\partial \mathcal{L}_k}{\partial w_{ij}}\bigg|_{w_{ij}=[w_{ij}]_l} \tag{3.212}$$

According to the chain rule, the derivative of the likelihood function, when considering the parameters a_{ij}, can be decomposed into:

$$\frac{\partial \mathcal{L}_k}{\partial w_{ij}^a} = \frac{\partial \mathcal{L}_k}{\partial a_{ij}} \frac{\partial a_{ij}}{\partial w_{ij}^a} \tag{3.213}$$

The leftmost derivative, at the right hand term, is already known from the previous section. The rightmost partial derivative can be computed according to:

$$
\begin{aligned}
\frac{\partial a_{ij}}{\partial w_{ij}^a} &= \frac{e^{w_{ij}^a} \sum_{l=1}^m e^{w_{il}^a} - e^{w_{ij}^a} \frac{\partial}{\partial w_{ij}^a}\left(e^{w_{i1}^a} + \cdots + e^{w_{ij}^a} + \cdots e^{w_{im}^a}\right)}{\left(\sum_{l=1}^m e^{w_l^a}\right)^2} \\
&= \frac{e^{w_{ij}^a} \sum_{k=1}^m e^{w_{ik}^a} - e^{w_{ij}^a} e^{w_{ij}^a}}{\left(\sum_{l=1}^m e^{w_{il}^a}\right)^2} \\
&= \frac{e^{w_{ij}^a}}{\sum_{l=1}^m e^{w_{il}^a}} - \left(\frac{e^{w_{ij}^a}}{\sum_{l=1}^m e^{w_{il}^a}}\right)^2 = a_{ij}\left(1 - a_{ij}\right)
\end{aligned} \tag{3.214}
$$

The same approach is used for the re-parameterization of b_{ij} and c_i leading to:

$$\frac{\partial b_{ij}}{\partial w_{ij}^b} = b_{ij}\left(1 - b_{ij}\right) \tag{3.215}$$

$$\frac{\partial c_i}{\partial w_i^c} = c_i\left(1 - c_i\right) \tag{3.216}$$

According to those final results, it is possible to describe an alternative learning method that maximizes the likelihood while, simultaneously, preserves the parameters stochasticity constraint. In this reference frame, the sequence of operations to be iteratively executed are presented in Algorithm 7.

The MATLAB® function described in Listing 3.22 can be used to estimate the model parameters using this re-parameterization method. The plot presented in Figure 3.17 was obtained using the above referred MATLAB® function with the training data generated from the execution

Algorithm 7 Training algorithm, according to the gradient rise, by re-parameterization of the transition probabilities.

1   Start from an initial admissible solution $\mathbf{A}$, $\mathbf{B}$ and $\mathbf{I}$ and a training sequence.
2   For $i,j = 1$ to m,
   2.1   calculate $\bar{\Omega}_a(i,j)$
   2.2 $w_{ij}^a = w_{ij}^a + \eta_A \cdot \bar{\Omega}_a(i,j) \cdot a_{ij}(1 - a_{ij})$
3   $a_{ij} = \dfrac{e^{w_{ij}^a}}{\sum_{l=1}^{m} e^{w_{il}^a}}$
4   For $i = 1$ to m and $j = 1$ to n,
   4.1   calculate $\bar{\Omega}_b(i,j)$
   4.2 $w_{ij}^b = w_{ij}^b + \eta_B \cdot \bar{\Omega}_b(i,j) \cdot b_{ij}(1 - b_{ij})$
5   $b_{ij} = \dfrac{e^{w_{ij}^b}}{\sum_{l=1}^{m} e^{w_{il}^b}}$
6   For $i = 1$ to m
   6.1 $w_i^c = w_i^c + \eta_c b_\xi(\lambda_1)$
7   $c_i = \dfrac{e^{w_i^c}}{\sum_{l=1}^{m} e^{w_i^c}}$
8   Repeat point (2) until the stop criterion is satisfied.

of the source code from Listing 3.19 and considering m = 2, n = 3, eta = [0.0010.0010.001] and MaxIter = 100.

Listing 3.22 Hidden Markov model estimation procedure using the gradient method by means of the re-parameterization procedure.

```
1  function  likelihood=test_gradient_reparam(O,m,n,eta,MaxIter)
2  % model characteristics...
3  % m- number of hidden states
4  % n- number of observable states
5  % eta - learning rate (vector)
6  % MaxIter - maximum number of iterations
7
8  taA=eta(1); etaB=eta(2); etac=eta(3); % learning rate
9  % ......................PARAMETERS INITIALIZATION
10 W=randn(m,m);A=exp(W)./((sum(exp(W).'))'*ones(1,m));
11 Q=randn(m,n);B=exp(Q)./((sum(exp(Q).'))'*ones(1,n));
12 Y=randn(m,1);c=exp(Y)./(sum(exp(Y)));
13 likelihood=zeros(1,MaxIter);
14
15 for j=1:MaxIter
16     [~,OMEGA]=gradientLA_norm(A,B,O,c);
17     dAdW=A.*(1-A);
18     W=W+etaA*OMEGA.*dAdW;
19     [~,OMEGA]=gradientLB_norm(A,B,O,c);
20     dBdW=B.*(1-B);
21     Q=Q+etaB*OMEGA.*dBdW;
22     [dLdc]=gradientLc(B,O);
23     dcdW=c.*(1-c);
24     Y=Y+etac*dLdc.*dcdW;
25     A=exp(W)./((sum(exp(W).'))'*ones(1,m));
26     B=exp(Q)./((sum(exp(Q).'))'*ones(1,n));
27     c=exp(Y)./(sum(exp(Y)));
```

```
28      ['',lik]=forward_algorithm_norm(A,B,O,c);
29      disp(['Iteration -- ' num2str(j) ' FIT -- ' num2str(lik)]);
30      likelihood(j)=lik;
31 end
```

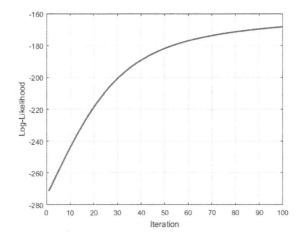

Figure 3.17 Evolution of the performance of the Markov model, over 100 iterations, evaluated according to each of the parameters of the model independently.

3.5.5 ROSEN'S ALGORITHM

In the previous section, we have assumed that the transition probability between states could be iteratively determined by means of a gradient ascend method. For example, the hidden states transition probability estimation equation, at iteration $l + 1$, was given by,

$$[\hat{a}_{ij}]_{l+1} = [\hat{a}_{ij}]_l + \eta \left.\frac{\partial \mathcal{L}_k}{\partial a_{ij}}\right|_{a_{ij}=[\hat{a}_{ij}]_l} \qquad (3.217)$$

At each iteration, the current solution $\hat{a}_{ij}$ will follow a direction, along the parameters space, is order to maximize the likelihood $\mathcal{L}_k$. The same can be said about the remaining model parameters such as $\hat{b}_{ij}$ and $\hat{c}_i$.

In a more general frame of reference, and considering only the model's hidden states layer, the estimation problem can be formulated as the following constrained optimization problem:

$$\max_{\mathbf{A}} \quad \mathcal{L}_k$$

$$s.t. \quad 0 \leq a_{ij} \leq 1$$

$$\sum_{k=1}^{m} a_{ik} = 1 \qquad (3.218)$$

$$i = 1, \cdots, m$$

$$j = 1, \cdots, m$$

where, the first constraint, represents the fact that the transition probabilities between states cannot be negative or greater than unity. Additionally, since for a given hidden state at time k, the transition probability from that state to another state, must be 1, then the second constraint must be added.

Notice that the problem cannot be solved by simply running equation (3.217) since the gradient is not free to move in all directions of the parameter space. It is confined to movements defined within the admissible region. That is, the spatial region in which all constraints of the problem are verified. For example, imagine the hypothetical situation illustrated in Figure 3.18.

The shaded region in the horizontal plane refers to the first constraint of (3.218) and the oblique line to the second constraint of the same expression. The admissible region is the intersection of the two constraints. This implies that the gradient of the objective function $\mathcal{L}_k$ can only be evaluated along the line drawn on the likelihood surface. Thus, (3.217) must be changed to:

$$[\hat{a}_{ij}]_{l+1} = [\hat{a}_{ij}]_l + \eta [d_{ij}]_l \qquad (3.219)$$

Because the search direction may not be given directly by the gradient $\frac{\partial \mathcal{L}_k}{\partial a_{ij}}$, to compute $[\hat{a}_{ij}]$, at iteration $l + 1$, it is necessary to obtain the effective search direction as well as the jump length defined by η. The magnitude of this jump can be dynamically changed at each iteration l from the condition that $[\hat{a}_{ij}]_{k+1}$ must be admissible. On the other hand, the search direction can only be selected among the universe of admissible directions. An admissible direction is one for which a small disturbance does not bring the solution out of the admissible region. For obvious reasons, and from the set of possible directions, the one most aligned with the gradient direction is the one that should be selected. That is, the one that

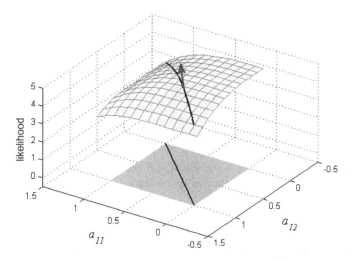

Figure 3.18 Objective function and admissible region (oblique line). On the objective function is represented the vector gradient along the projection of the admissible region on the objective function.

makes the tightest angle with the direction of $\frac{\partial \mathcal{L}_k}{\partial a_{ij}}$. Figure 3.19 illustrates this idea. Of the three possible directions presented, the direction to follow by the search algorithm must be d_3.

The idea of using the search direction pointed out by the vector resulting from the projection of the gradient into the admissible region, was introduced by Rosen in the early 1960s [27]. The Rosen's algorithm, also known as the gradient projection method, starts from an admissible initial solution and, by pursuing successive admissible directions, will gradually obtain solutions that have a better performance regarding the objective function. At the end of the iterations process, the present method ensures that the optimum found is within, or at the boundary, of the admissible region. Notice that, unless the optimization problem is convex, there is no guarantee that the optimum attained is the global optimum and not just a local optimum.

The Rosen algorithm can be applied to a large class of optimization problems. However, in this book, we are only concerned with its use over the hidden Markov model parameter estimation problem. Nonetheless, before using it within this framework, Rosen's method will be first addressed generically considering the situation of having both equality and inequality constraints.

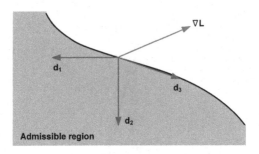

Figure 3.19 Representation of the gradient and three other possible admissible directions d_1, d_2 and d_3. The latter vector is the one whose projection along the gradient is larger.

3.5.5.1 Linear equality constraints

The first situation to be evaluated assumes the minimization of a multivariate function $f(\mathbf{x})$ subject to a set of equality constraints. The problem is formulated as follows:

$$\min_{\mathbf{x}} \quad f(\mathbf{x})$$
$$s.t. \quad \mathbf{A}\mathbf{x} - \mathbf{b} = \mathbf{0} \tag{3.220}$$

where $\mathbf{A}$ is an array with dimension $r \times n$. The variable r refers to the number of constraints and n to the dimension of $\mathbf{x}$. Additionally, $\mathbf{b}$ and $\mathbf{0}$ are column vectors with dimension r.

Let's start by assuming an admissible solution $\mathbf{x}_l$ such that,

$$\mathbf{A}\mathbf{x}_l - \mathbf{b} = \mathbf{0} \tag{3.221}$$

Based on this solution, and since we are dealing with a minimization problem, the objective is to find another admissible one, $\mathbf{x}_{l+1}$, such that,

$$f(\mathbf{x}_l) < f(\mathbf{x}_{l+1})$$

The solution $\mathbf{x}_{l+1}$ will be obtained from $\mathbf{x}_l$ using:

$$\mathbf{x}_{l+1} = \mathbf{x}_l + \eta_l \cdot \mathbf{d}_l \tag{3.222}$$

where $\mathbf{d}$ refers to a direction vector with unity norm and η the distance between the two consecutive solutions.

The question that remains is the choice of the direction and the length of the "jump". Regarding the former, a natural choice will be the one that leads to a higher rate of variation of $f(\mathbf{x})$ regarding a disturbance at $\mathbf{x}$. That is,

$$\mathbf{d}_l = \nabla f(\mathbf{x})\,|_{\mathbf{x}=\mathbf{x}_l}$$

However, in problems with constraints, not all the directions can be taken. Directions that push the solution to non-admissible regions are prohibited. Furthermore, even if the direction is valid, the value of η_l can bring the solution out of the admissible space. Therefore, it is necessary to establish, in an efficient way, both the direction and the value of the learning ratio in order to guarantee that, between iterations, the solution remains always admissible.

If $\mathbf{x}_{l+1}$ is admissible, and considering (3.222), one obtains:

$$\mathbf{A}\mathbf{x}_{l+1} - \mathbf{b} = \mathbf{0} \Leftrightarrow$$
$$\mathbf{A}(\mathbf{x}_l + \eta_l \cdot \mathbf{d}_l) - \mathbf{b} = \mathbf{0} \Leftrightarrow \qquad (3.223)$$
$$\mathbf{A}\mathbf{x}_l + \eta_l \cdot \mathbf{A}\mathbf{d}_l - \mathbf{b} = \mathbf{0}$$

Since $\mathbf{x}_l$ is admissible, then $\mathbf{A}\mathbf{x}_l - \mathbf{b} = \mathbf{0}$ and thus,

$$\eta_l \cdot \mathbf{A}\mathbf{d}_l = \mathbf{0} \underset{\eta>0}{\Rightarrow} \mathbf{A}\mathbf{d}_l = \mathbf{0} \qquad (3.224)$$

In addition, since the norm of the vector $\mathbf{d}$ is one, then,

$$\mathbf{d}_l^T \mathbf{d}_l = 1 \qquad (3.225)$$

The idea is to minimize the projection of the gradient over the direction $\mathbf{d}_l$. It is known that the opposite direction to the one pointed out by the gradient, is that which, locally, indicates the direction where the function variation is larger. However, as already referred, not all directions are admissible. Of all the possible admissible directions, we seek for the one that has the larger collinearity possible with the opposite direction pointed out by the gradient vector (without forgetting the restrictions imposed to $\mathbf{d}_l$). If $\nabla^T f(\mathbf{x}_l) \cdot \mathbf{d}_l$ is the projection of the vector $\mathbf{d}_l$ over the gradient vector, then $\mathbf{d}_l$ must be chosen as the solution of the following optimization problem:

$$\min_{\mathbf{d}_l} \quad \nabla^T f(\mathbf{x}_l) \cdot \mathbf{d}_l$$
$$s.t. \quad \mathbf{A}\mathbf{d}_l = \mathbf{0} \qquad (3.226)$$
$$\mathbf{d}_l^T \mathbf{d}_l - 1 = 0$$

In order to solve it, the method of Lagrange multipliers can be applied. The Lagrange multipliers is a technique for finding the local maxima and minima of a function subject to equality constraints. In order to do this, first the Lagrangian function must be formed. For the above optimization problem, the Lagrangian takes the following shape:

$$\mathbf{L}\left(\mathbf{d}_l, \boldsymbol{\lambda}, \lambda_0\right) = \nabla^T f\left(\mathbf{x}_l\right) \cdot \mathbf{d}_l + \boldsymbol{\lambda}^T \mathbf{A} \mathbf{d}_l + \lambda_0 \left(\mathbf{d}_l^T \mathbf{d}_l - 1\right) \tag{3.227}$$

where $\boldsymbol{\lambda}$ and λ_0 are the Lagrange multipliers. Notice that the former is a vector with dimension $r \times 1$ and the latter a scalar.

Finding the critical point involves the solution of the following system of equations:

$$\frac{\partial \mathbf{L}\left(\mathbf{d}_l, \boldsymbol{\lambda}, \lambda_0\right)}{\partial \mathbf{d}_l} = \nabla f\left(\mathbf{x}_l\right) + \mathbf{A}^T \boldsymbol{\lambda} + 2\lambda_0 \mathbf{d}_l = 0 \tag{3.228a}$$

$$\frac{\partial \mathbf{L}\left(\mathbf{d}_l, \boldsymbol{\lambda}, \lambda_0\right)}{\partial \boldsymbol{\lambda}} = \mathbf{A} \mathbf{d}_l = 0 \tag{3.228b}$$

$$\frac{\partial \mathbf{L}\left(\mathbf{d}_l, \boldsymbol{\lambda}, \lambda_0\right)}{\partial \lambda_0} = \mathbf{d}_l^T \mathbf{d}_l - 1 = 0 \tag{3.228c}$$

Expressing (3.228a) in terms of $\mathbf{d}_l$ yields:

$$\mathbf{d}_l = -\frac{1}{2\lambda_0}\left(\nabla f\left(\mathbf{x}_l\right) + \mathbf{A}^T \boldsymbol{\lambda}\right) \tag{3.229}$$

Now, multiplying (3.228a) by $\mathbf{A}$, and taking into account the equality expressed by (3.228b), leads to:

$$\mathbf{A}\nabla f\left(\mathbf{x}_l\right) + \mathbf{A}\mathbf{A}^T \boldsymbol{\lambda} + 2\lambda_0 \mathbf{A}\mathbf{d}_l = 0$$
$$\Rightarrow \mathbf{A}\nabla f\left(\mathbf{x}_l\right) + \mathbf{A}\mathbf{A}^T \boldsymbol{\lambda} = 0 \tag{3.230}$$

which, after isolating $\boldsymbol{\lambda}$, and assuming that $\mathbf{A}\mathbf{A}^T$ is invertible[5], the following result is obtained:

$$\boldsymbol{\lambda} = -\left(\mathbf{A}\mathbf{A}^T\right)^{-1}\mathbf{A}\nabla f\left(\mathbf{x}_l\right) \tag{3.231}$$

Going back, and replacing this last equality into (3.229), yields:

$$\mathbf{d}_l = -\frac{1}{2\lambda_0}\left(\nabla f\left(\mathbf{x}_l\right) - \mathbf{A}^T\left(\mathbf{A}\mathbf{A}^T\right)^{-1}\mathbf{A}\nabla f\left(\mathbf{x}_l\right)\right) \tag{3.232}$$

---

[5]Which implies that all constraints equations are linearly independent.

which can be expressed, after factorization of $\nabla f(\mathbf{x}_l)$, as:

$$\mathbf{d}_l = \frac{1}{2\lambda_0}\left(\mathbf{A}^T\left(\mathbf{A}\mathbf{A}^T\right)^{-1}\mathbf{A} - \mathbf{I}\right)\nabla f(\mathbf{x}_l) \qquad (3.233)$$

The value of the Lagrange multiplier λ_0 is still unknown. Nonetheless, notice that (3.228c) can be expressed as $\|\mathbf{d}_l\| = 1$ and, by taking into consideration (3.233), then:

$$\|\mathbf{d}_l\| = \left\|\frac{1}{2\lambda_0}\left(\mathbf{A}^T\left(\mathbf{A}\mathbf{A}^T\right)^{-1}\mathbf{A} - \mathbf{I}\right)\nabla f(\mathbf{x}_l)\right\| = 1 \qquad (3.234)$$

Since $\frac{1}{2\lambda_0}$ is constant, the previous expression is formulated as[6]:

$$\lambda_0 = \frac{1}{2}\left\|\left(\mathbf{A}^T\left(\mathbf{A}\mathbf{A}^T\right)^{-1}\mathbf{A} - \mathbf{I}\right)\nabla f(\mathbf{x}_l)\right\| \qquad (3.235)$$

Finally, the algorithm search direction is obtained from[7]:

$$\mathbf{d}_l = \frac{\left(\mathbf{A}^T\left(\mathbf{A}\mathbf{A}^T\right)^{-1}\mathbf{A} - \mathbf{I}\right)\nabla f(\mathbf{x}_l)}{\left\|\left(\mathbf{A}^T\left(\mathbf{A}\mathbf{A}^T\right)^{-1}\mathbf{A} - \mathbf{I}\right)\nabla f(\mathbf{x}_l)\right\|} \qquad (3.236)$$

As can be seen, the resulting vector is truly unitary since the vector presented in the numerator divides by its norm. Additionally, the matrix obtained after the operation $\mathbf{A}^T\left(\mathbf{A}\mathbf{A}^T\right)^{-1}\mathbf{A} - \mathbf{I}$, is commonly referred to as the projection matrix.

Recall that, from (3.222), the motion direction of $\mathbf{x}_l$ is already known. What remains to be found is the step value η_l. Care must be taken to define this coefficient since a value too large can cause $\mathbf{x}_{l+1}$ to escape from the admissible region. Note also that, since the gradient only provides local information, a step too high can lead the solution to a location where the direction of the gradient is substantially different. In the limit, a value too high can lead to algorithm instability and lack of convergence.

For example, consider an optimization problem with two unknowns and one constraint. Geometrically, this constraint describes a straight line. Since the search direction refers to the projection of the gradient vector into the constraint space then, regardless of the value of η, the solution

---

[6]Notice that, if k is a constant, then $\|k\mathbf{x}\| = |k| \cdot \|x\|$.

[7]In practice it is necessary to take into account the situation where ∇f is zero because, in this case, the denominator of $\mathbf{d}_l$ is also zero.

x_{l+1} will always be admissible. Now, scaling up this example to the situation involving three decision variables with two constraints, then the admissible region is defined by the intersection of two planes. If the planes are parallel, and assuming that the constraints equations are linearly independent, there is no admissible solution. On the other hand, if they are not parallel, their intersection leads to a line of admissible solutions. Therefore, it boils down to a situation similar to the previous one. As a matter of fact, this observation can be extrapolated to a problem of any size. That is, for optimization problems with only linear equality type constraints there is no upper bound for η. In order to illustrate this statement, suppose that it is intended to find the solution that minimizes a quadratic type function subject to an equality constraint. For example, an optimization problem similar to the one provided below:

$$
\begin{aligned}
\min_{\mathbf{x}} \quad & \mathbf{x}^T \mathbf{x} + 1 \\
s.t. \quad & [3 \quad -6]\mathbf{x} - 3 = 0
\end{aligned}
\tag{3.237}
$$

The objective function geometric shape, as well as the constraint, is presented in Figure 3.20. Beginning with an arbitrary admissible solution $\mathbf{x}_0$, a better solution can be obtained by means of:

$$
\mathbf{x}_1 = \mathbf{x}_0 + \eta \cdot \mathbf{d}_0
\tag{3.238}
$$

where $\mathbf{x}_1$ is the new problem solution and $\mathbf{d}_0$ the search direction which must be computed. Let's consider $\mathbf{x}_0 = [3 \quad 1]^T$, as the initial solution. The objective function value at this point is equal to 11 and the search direction $\mathbf{d}_0$ will be obtained from (3.236). Hence, it is necessary to calculate both $\nabla f(\mathbf{x})$ and the projection matrix. Regarding the former[8], it is straightforward to see that $\nabla f(\mathbf{x}) = 2\mathbf{x}$.

On the other hand, the projection matrix is obtained as follows:

$$
\begin{aligned}
& \mathbf{A}^T \left(\mathbf{A}\mathbf{A}^T \right)^{-1} \mathbf{A} - \mathbf{I} = \\
& \begin{bmatrix} 3 \\ -6 \end{bmatrix} \begin{bmatrix} 9 & -18 \\ -18 & 36 \end{bmatrix}^{-1} \begin{bmatrix} 3 \\ -6 \end{bmatrix} - \begin{bmatrix} 1 & 0 \\ 0 & 1 \end{bmatrix} \\
& = - \begin{bmatrix} 0.8 & 0.4 \\ 0.4 & 0.2 \end{bmatrix}
\end{aligned}
\tag{3.239}
$$

---

[8]Note that $\frac{d\mathbf{x}^T \mathbf{B}\mathbf{x}}{d\mathbf{x}} = \left(\mathbf{B} + \mathbf{B}^T \right) \mathbf{x}$

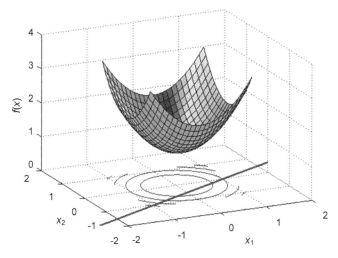

Figure 3.20 Quadratic function subject to an equality constraint as defined by the expression (3.237).

The numerator of (3.236) will then become the vector $[-5.6 \ -2.8]^T$ which, after normalization, leads to:

$$\mathbf{d}_0 \approx - \left[\begin{array}{c} 0.8944 \\ 0.4472 \end{array} \right] \tag{3.240}$$

Assuming $\eta = 1.5$, the new solution becomes:

$$\mathbf{x}_1 = \left[\begin{array}{c} -1.0249 \\ -1.0124 \end{array} \right] \tag{3.241}$$

For this new solution, the objective function value is approximately 3. Which is a value substantially smaller than the one provided by the initial solution $\mathbf{x}_0$.

Figure 3.21 graphically presents the vectors $-\nabla f(\mathbf{x})$ and $\mathbf{d}_0$ associated to the initial point $\mathbf{x}_0$. Notice that, regardless of the value of η, the next solution $\mathbf{x}_1$ will always be admissible since it will always be somewhere over the line defined by the constraint.

Although the value of η does not interfere with the admissibility of the new solution, it has a major impact on the algorithm performance. To illustrate this fact, Table 3.6 shows the performance of the solution for three different values of η. For $\eta = 0.1$, the performance of $\mathbf{x}_1$ increases

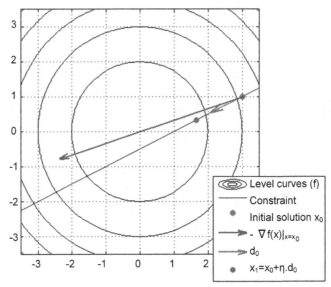

Figure 3.21 Initial direction of the gradient vector $\nabla f(x_0)$ and its projection, $\mathbf{d}_0$ over the admissible region.

Color version at the end of the book

around 5.6% comparatively to $\mathbf{x}_0$. The performance is further increased if $\eta = 1$ but decreases dramatically for $\eta = 10$.

The algorithm sensitivity to η can be further noticed from Figure 3.22 where the convergence plots, taken for 50 iterations, are presented for four distinct values of η. Remark that, if the value of η is too large, the algorithm will become unstable. On the other hand, if it is too small, the algorithm convergence will be very slow.

Table 3.6

Performance of $\mathbf{x}_1$ as a function of the learning rate η.

	Learning rate		
	$\eta = 0.1$	$\eta = 1$	$\eta = 10$
$f(\mathbf{x}_1)$	10.384	5.739	48.39
$100\frac{\Delta f(\mathbf{x}_1)}{f(\mathbf{x}_0)}$	5.6%	47.8%	-340%

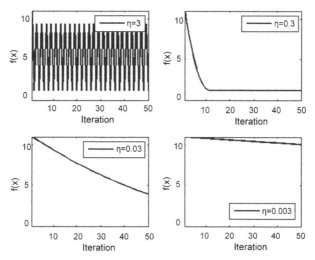

Figure 3.22 Convergence of the gradient projection algorithm considering four different values for η.

3.5.5.2 Lagrange multipliers and Karush-Kuhn-Tucker conditions

At this moment, it is important to discuss the necessary conditions that must be met by a constrained optimization problem to guarantee the existence of a solution.

For an arbitrary vectorial function $f(\mathbf{x})$, its gradient, denoted by $\nabla f(\mathbf{x})$, will be a vector whose components are the partial derivatives of $f(\mathbf{x})$ in order to each of the vector components. That is, if $\mathbf{x} = [x_1 \quad \cdots \quad x_n]$ then,

$$\nabla f(\mathbf{x}) = \left[\frac{\partial f(\mathbf{x})}{\partial x_1} \quad \cdots \quad \frac{\partial f(\mathbf{x})}{\partial x_n}\right] \tag{3.242}$$

To illustrate this, and by considering the parametric equation $f(x) = x^2$, then $\nabla f(x) = \frac{\partial f(x)}{\partial x} = 2x$. For an arbitrary point over the curve $f(x)$, say $x = 2$, the gradient function at this point is therefore equal to 4.0. Figure 3.23 illustrates the shape of $f(x)$ as well as the gradient vector at the above referred point.

The gradient vector has the particularity of indicating, locally, what is the function increment direction. From the previous figure, it is possible to conclude that, around $x = 2$, moving the location of x to the right will lead to an increase on the function value.

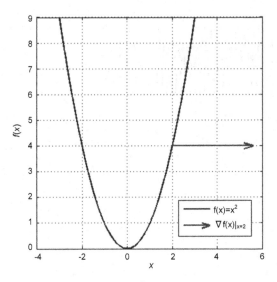

Figure 3.23 Quadratic function $f(x) = x^2$ and its vector gradient at $x = 2$.

Consider now a two dimensional function with the following formulation:

$$f(x_1, x_2) = x_1^3 - 3x_1 + x_2^2 + 5 \qquad (3.243)$$

Assuming $-3 \leq x_1 \leq 3$ and $-4 \leq x_2 \leq 4$, this function has the shape illustrated at Figure 3.24 and its contour line plot can be found at Figure 3.25. Along the contour lines, the value of the function is always equal. Moreover, the area between the curves is filled with a color code where higher values of $f(x_1, x_2)$ correspond to lighter colors.

The gradient is $\nabla f(x_1, x_2) = \begin{bmatrix} 3x_1^2 - 3 & 2x_2 \end{bmatrix}$ which, for particular pairs of (x_1, x_2), such, $(1.5, 1)$, $(0.5, -0.5)$, $(1.6, -1.2)$ and $(-1, 1)$ leads to the graph representation of Figure 3.26.

As can be seen, the gradient vector at a given point over the level curve is perpendicular to the tangent line to the curve at that point. This is a very important consideration that must be kept in mind and will be used later on.

It can be seen that there is a close relationship between the tangent slope on a point, over the level curve, and the gradient vector at that point. The line equation, tangent to the function $x_2 = f(x_1)$, at any arbitrary point (x_1', x_2') is given by,

$$(x_2 - x_2') = -m (x_1 - x_1') \qquad (3.244)$$

118

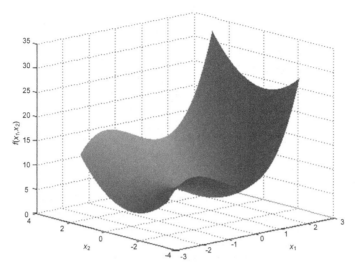

Figure 3.24 Spatial representation of the function $f(x_1, x_2) = x_1{}^3 - 3x_1 + x_2{}^2 + 5$.

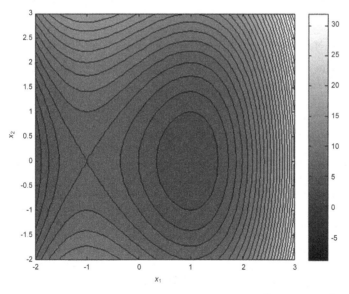

Figure 3.25 Contours lines for the function $f(x_1, x_2) = x_1{}^3 - 3x_1 + x_2{}^2 + 5$.

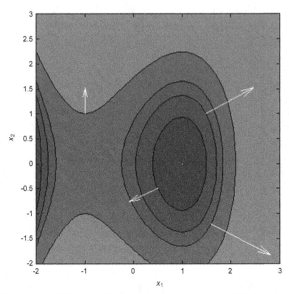

Figure 3.26 Direction of the gradient vector in several points of the function 3.243 surface.

where the slope is $m = \frac{\partial x_2}{\partial x_1}\Big|_{x_1=x_1'}$.

Using the derivative chain rule, it is possible to express the line slope at the point (x_1', x_2') as:

$$m = \frac{\partial f(x_1, x_2)\partial x_2}{\partial x_1 \partial f(x_1, x_2)} \tag{3.245}$$

where $f(x_1, x_2)$ refers to the level curve equation. Thus, (3.244) can now be represented as:

$$\frac{\partial f(x_1, x_2)}{\partial x_2}(x_2 - x_2') = -\frac{\partial f(x_1, x_2)}{\partial x_1}(x_1 - x_1') \tag{3.246}$$

The previous equation can be expressed using matricial notation as:

$$\nabla f(x_1, x_2)|_{x_1=x_1', x_2=x_2'} \cdot \mathbf{x} = \nabla f(x_1, x_2)|_{x_1=x_1', x_2=x_2'} \cdot \mathbf{x'} \tag{3.247}$$

where $\mathbf{x} = [x_1 \quad x_2]^T$ and $\mathbf{x'} = [x_1' \quad x_2']^T$.

Remark that, in any $\mathbf{A}\mathbf{x} = b$ formulation, each vector in $\mathbf{A}$ is normal to the defined hyperplane. Hence, in (3.247), it is easy to see that the gradient vector is normal to the tangent at any point over the level curve. Moreover, it is also known that in a function local extremum, the slope of the tangent line at that point is zero. That is, the gradient at this point is equal to the null vector. The points at which this phenomenon happens are called the critical points of a function. For example, for the curve $f(x) = x^2$, $\nabla f(x) = 2x$. Equating this last expression to zero and solving for x, we get the unique solution $x = 0$. That is, the function has an extremum at $x = 0$. Notice that it is not possible to say that this extremum point is a maximum or a minimum of the function. In order to be able to determine if the extremum point is a maximum or a minimum, the information provided by the Hessian matrix is required.

If $f(\mathbf{x})$, for $\mathbf{x} = [x_1 \cdots x_n]$, is a differentiable function (at least up to order two), the Hessian matrix has the following structure:

$$H\{f(\mathbf{x})\} = \begin{bmatrix} \frac{\partial^2 f(\mathbf{x})}{\partial x_1^2} & \cdots & \frac{\partial^2 f(\mathbf{x})}{\partial x_1 \partial_n} \\ \vdots & \ddots & \vdots \\ \frac{\partial^2 f(\mathbf{x})}{\partial x_n \partial x_1} & \cdots & \frac{\partial^2 f(\mathbf{x})}{\partial x_n^2} \end{bmatrix} \tag{3.248}$$

Now, if $\mathbf{x} = \mathbf{x}^*$ is a critical point of $f(\mathbf{x})$ and if the Hessian matrix is positive definite, then the point $\mathbf{x}^*$ is a local minimum. If, on the other hand, the Hessian is negative definite, then the point $\mathbf{x} = \mathbf{x}^*$ refers to a local maximum. A matrix is said to be positive definite if all its eigenvalues[9] are positive. If all of them are negative, the matrix is said to be negative definite. It is also possible to consider positive semi-definite and negative semi-definite matrices. In these cases, there are some eigenvalues that are zero. A matrix is undefined if there are both positive and negative eigenvalues. Notice that a necessary condition for the convexity of a function is that its Hessian is positive (semi) definite.

From $f(x) = x^2$, it was seen that $x = 0$ was a critical point but there was no information if that extremum refers to a maximum or a minimum. The second derivative of $f(x)$ at $x = 0$ is equal to:

$$\frac{\partial^2 f(x)}{\partial x^2}\bigg|_{x=0} = 2 > 0 \tag{3.249}$$

so the point $x = 0$ refers to a minimum.

---

[9]The eigenvalues of a matrix refer to the solutions of the characteristic equation $(\mathbf{A} - \lambda\mathbf{I})\mathbf{x} = \mathbf{0}$, for $\mathbf{x} \neq \mathbf{0}$. In order for this equality to occur it is necessary that the matrix $(\mathbf{A} - \lambda\mathbf{I})$ is non invertible. That is, $\det(\mathbf{A} - \lambda\mathbf{I}) = 0$. The λ values that validate this equality are called eigenvalues.

Returning to (3.243), in this case the critical points are obtained from solving $\nabla f(x_1, x_2) = \mathbf{0}$. That is,

$$\frac{\partial f(x_1, x_2)}{\partial x_1} = 3x_1{}^2 - 3 = 0$$
$$\frac{\partial f(x_1, x_2)}{\partial x_2} = 2x_2 = 0 \tag{3.250}$$

leading to $x_1 = \pm 1$ and $x_2 = 0$.

Thus, there are two different critical points for the function $f(x_1, x_2)$ located at $(-1, 0)$ and $(1, 0)$. To obtain more information about these extremes, the Hessian matrix is calculated:

$$H\{f(x_1, x_2)\} = \begin{bmatrix} 6x_1 & 0 \\ 0 & 2 \end{bmatrix} \tag{3.251}$$

Since the Hessian is a diagonal matrix, then its eigenvalues are the elements of its main diagonal. In this case, $6x_1$ and 2.

For the critical point $(1,0)$, the eigenvalues are 6 and 2, both positive. Thus the matrix is positive definite and the point in question refers to a minimum. On the other hand, for $(-1, 0)$, the eigenvalues are -6 and 2 which implies that the Hessian is indefinite. As a matter of fact, by looking at the Figures 3.23 and 3.24 it is possible to confirm that this point refers to a "saddle" point. That is, there are directions departing from that point where the function increases in value and there are other directions for which the function value decreases.

The computation of the extremum of a function $f(\mathbf{x})$, from the analysis of its critical points, is limited to situations in which the independent variables are inter-independent. In other words, the value of a variable does not affect the value of the remaining variables. However, for functions whose independent variables are subject to constraints, this is not true. The use of constraints implies that the variables set cease to be independent and become dependent. The relationship between these variables can be expressed by a set of equations of the form:

$$g_1(\mathbf{x}) = c_1$$
$$\cdots \tag{3.252}$$
$$g_k(\mathbf{x}) = c_k$$

where c_i for $i = 1, \cdots, k$ are constants. In this situation, the optimization problem, subject to those set of constraints, is formulated as:

$$\max_{\mathbf{x}} / \min_{\mathbf{x}} \quad f(\mathbf{x})$$
$$s.t. \quad g_1(\mathbf{x}) = c_1$$
$$\ldots$$
$$g_k(\mathbf{x}) = c_k \tag{3.253}$$

The above gradient based method can not be used to find the critical point of $f(\mathbf{x})$ in this situation. The reason is that the extremum point obtained by this method may not validate all the constraints. In the case of constrained optimization, other techniques must be used. For example the Lagrange multipliers method.

The Lagrangian multipliers method targets the optimization of a differentiable functions subject to constraints also differentiable. It is important to refer that both objective function and constraints can be nonlinear. The Lagrange multipliers method can handle constraints of both equality and inequality type. Throughout this section, both situations will be analyzed.

To illustrate this method, let's return to (3.243) and assume the addition of the constraint $9x_1 + 2x_2 = 20$. The optimization problem is then formulated as:

$$\min_{\mathbf{x}} \quad f(x_1, x_2) = x_1{}^3 - 3x_1 + x_2{}^2 + 5$$
$$s.t. \quad g(x_1, x_2) = 9x_1 + 2x_2 - 20 = 0 \tag{3.254}$$

The reader can gain intuition regarding the above problem by observing Figure 3.27 where the shape of both the objective function and constraints is presented.

The solution to the above problem is the point at which the line,

$$g(x_1, x_2) = 20$$

is tangent to the level curve of $f(x_1, x_2) = f(x_1^*, x_2^*)$. The pair (x_1^*, x_2^*) refers to the optimal point of the problem.

It is known that $\nabla f(x_1, x_2)|_{x_1=x_1^*, x_2=x_2^*}$ is a vector orthogonal to the level curve at the point (x_1^*, x_2^*). It is also clear that $\nabla g(x_1, x_2)$ is a vector normal to the line defined by $g(x_1, x_2) = 0$. Thus, at the point of intersection of the curves $f(x_1, x_2)$ and $g(x_1, x_2)$, (x_1^*, x_2^*), the vector $\nabla f(x_1, x_2)$ must have the same direction as the vector $\nabla g(x_1, x_2)$. Mathematically this condition can be expressed as:

$$\nabla f(x_1, x_2) = \lambda \cdot \nabla g(x_1, x_2) \tag{3.255}$$

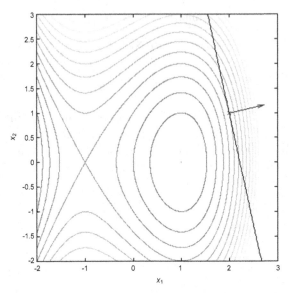

Figure 3.27 Contour lines and constraint function established in (3.254).

where λ refers to a proportionality constant called the Lagrange multiplier. Note that, if $\lambda < 0$, the vectors have opposite directions. On the other hand, if $\lambda = 0$, then $\nabla f(x_1, x_2) = 0$ which implies that the extremum of $f(x_1, x_2)$ also verifies the constraint $g(x_1, x_2)$. Since $\nabla f(x_1, x_2) = \begin{bmatrix} 3x_1^2 - 3 & 2x_2 \end{bmatrix}$ and $\nabla g(x_1, x_2) = \begin{bmatrix} 9 & 2 \end{bmatrix}$, applying the condition (3.255) gives the following system of equations:

$$\begin{cases} 3x_1^2 - 3 = 9\lambda \\ 2x_2 = 2\lambda \end{cases} \tag{3.256}$$

Moreover, the constraint $9x_1 + 2x_2 = 20$ must be verified which leads to two possible solutions: $x_1 = 2$, $x_2 = 1$ and $\lambda = 1$ in the first case and $x_1 = -15.5$, $x_2 = 79.75$ and $\lambda = 79.75$ in the second.

It is worth to notice that the use of the Lagrange multipliers technique only ensures that the solutions found are critical points. There is nothing throughout the problem resolution that indicates whether the solution concerns a maximum or a minimum. This information can be retrieved by inspecting the Hessian. However, in the case of constrained problems, the Hessian is generated differently and is therefore termed "bordered Hessian". As an example, for a n variable optimization problem with m constraints, the bordered Hessian associated with this problem leads to a square matrix with dimension $(n + m) \times (n + m)$. This matrix has the following structure:

$$H_b = \begin{bmatrix} \mathbf{0} & G \\ G^T & H \end{bmatrix} \tag{3.257}$$

where H refers to the conventional objective function Hessian and G is a $m \times n$ matrix of constraints partial derivatives. For example, for the problem:

$$\begin{aligned} \max_{\mathbf{x}} \quad & f(x_1, x_2, \cdots, x_n) \\ s.t. \quad & g(x_1, x_2, \cdots x_n) = 0 \end{aligned} \tag{3.258}$$

the bordered Hessian is:

$$H_b\{f \mid g\} = \begin{bmatrix} 0 & \frac{\partial g}{\partial x_1} & \frac{\partial g}{\partial x_2} & \cdots & \frac{\partial g}{\partial x_n} \\ \frac{\partial g}{\partial x_1} & \frac{\partial^2 f}{\partial x_1^2} & \frac{\partial^2 f}{\partial x_1 \partial x_2} & \cdots & \frac{\partial^2 f}{\partial x_1 \partial x_n} \\ \frac{\partial g}{\partial x_2} & \frac{\partial^2 f}{\partial x_2 \partial x_1} & \frac{\partial^2 f}{\partial x_2^2} & \cdots & \frac{\partial^2 f}{\partial x_2 \partial x_n} \\ \vdots & \vdots & \vdots & \ddots & \vdots \\ \frac{\partial g}{\partial x_n} & \frac{\partial^2 f}{\partial x_n \partial x_1} & \frac{\partial^2 f}{\partial x_n \partial x_2} & \cdots & \frac{\partial^2 f}{\partial x_n^2} \end{bmatrix} \tag{3.259}$$

If $\{\det(H_b)\} = (-1)^n$, then the problem extremum is a local maximum. Alternatively, if $\{\det(H_m)\} = (-1)^m$ it refers to a local minimum.

In short, the Lagrange multipliers method consists on solving the system of equations established by:

$$\begin{aligned} \nabla f(\mathbf{x}) &= \lambda \nabla g(\mathbf{x}) \\ g(\mathbf{x}) &= 0 \end{aligned} \tag{3.260}$$

For the situations that involve multiple constraint equations, the gradient of $f(\mathbf{x})$ must be parallel to the gradient of *all* existing constraints. That is, for a number n of constraints, the system of equations (3.260) becomes:

$$\begin{aligned} \nabla f(\mathbf{x}) &= \lambda_1 \nabla g_1(\mathbf{x}) \\ \nabla f(\mathbf{x}) &= \lambda_2 \nabla g_2(\mathbf{x}) \\ &\cdots \\ \nabla f(\mathbf{x}) &= \lambda_n \nabla g_n(\mathbf{x}) \\ g_1(\mathbf{x}) &= 0 \\ g_2(\mathbf{x}) &= 0 \\ &\cdots \\ g_n(\mathbf{x}) &= 0 \end{aligned} \tag{3.261}$$

All of these conditions can be fused into a much more compact formulation by defining the Lagrangian function:

$$L(\mathbf{x}, \lambda_1, \cdots, \lambda_n) = f(\mathbf{x}) + \sum_{i=1}^{n} \lambda_i \cdot g_i(\mathbf{x}) \qquad (3.262)$$

The relationship between this last expression and the previous one may not be evident. However, notice for example that $\frac{\partial L(\mathbf{x}, \lambda)}{\partial \lambda_i} = 0$, converts to $g_i(\mathbf{x}) = 0$ which is the equation associated to the *ith* constraint.

Defining $\lambda = [\lambda_1 \quad \cdots \quad \lambda_n]^T$ and $\mathbf{g}(\mathbf{x}) = [g_1(\mathbf{x}) \quad \cdots \quad g_n(\mathbf{x})]^T$, the above expression can be represented in matricial form by:

$$L(\mathbf{x}, \lambda) = f(\mathbf{x}) + \lambda^T \mathbf{g}(\mathbf{x}) \qquad (3.263)$$

In this case, the critical points are obtained by solving:

$$\nabla L(\mathbf{x}, \lambda) = \mathbf{0} \qquad (3.264)$$

which, as already mentioned, is nothing more than a compact way of expressing all the constraints in (3.261).

Now, consider the situation with inequality constraints in the problem formulation. In particular, consider an optimization problem, possibly non-linear but differentiable, as follows:

$$\begin{aligned} &\min_{\mathbf{x}} \quad f(\mathbf{x}) \\ &s.t. \quad g_i(\mathbf{x}) = 0 \\ &\qquad h_j(\mathbf{x}) \leq 0 \end{aligned} \qquad (3.265)$$

with $i = 1, \cdots, p$ and $j = 1, \cdots, q$. In this case, the variable p refers to the number of equality constraints and q to the number of inequality constraints. It is also assumed that the functions $f(\mathbf{x})$, $g_i(\mathbf{x})$ and $h_j(\mathbf{x})$ are, at least, twice differentiable.

The inequality constraints can be accommodated, in the Lagrange multipliers reference frame, by the addition of auxiliary variables denominated by slack variables. Each inequality constraint will be associated to a slack variable $\xi^2 \geq 0$ so that the inequality becomes an equality constraint[10]. Thus, the above problem can be framed as a regular mini-

---

[10]The square in ξ allows to guarantee that, regardless of the value of ξ, ξ^2 is always positive or zero.

mization problem subject only to equality constraints with form:

$$\min_{\mathbf{x}} \quad f(\mathbf{x})$$
$$s.t. \quad g_i(\mathbf{x}) = 0$$
$$h_j(\mathbf{x}) + \xi_j^2 = 0 \qquad (3.266)$$
$$\xi_j^2 \geq 0$$

The Lagrangian associated with this problem, has the following formulation:

$$L(\mathbf{x}, \boldsymbol{\lambda}, \boldsymbol{\mu}, \boldsymbol{\phi}) = f(\mathbf{x}) + \sum_{i=1}^{p} \lambda_i \cdot g_i(\mathbf{x}) + \sum_{i=1}^{q} \mu_i \cdot \left(h_i(\mathbf{x}) + \xi_i^2\right) \qquad (3.267)$$

where $\boldsymbol{\lambda} = [\lambda_1 \quad \cdots \quad \lambda_p]$ and $\boldsymbol{\mu} = [\mu_1 \quad \cdots \quad \mu_q]$ refer to the Lagrange multipliers associated with the equality and inequality constraints respectively. In addition, $\boldsymbol{\xi} = [\xi_1 \quad \cdots \quad \xi_q]$ defines a vector whose components are the slack variables. Solving for $\nabla L(\mathbf{x}, \boldsymbol{\lambda}, \boldsymbol{\mu}, \boldsymbol{\xi}) = \mathbf{0}$, leads to the following sequence of equations:

$$\nabla_{\mathbf{x}} L(\mathbf{x}, \boldsymbol{\lambda}, \boldsymbol{\mu}, \boldsymbol{\xi}) = \qquad (3.268a)$$

$$\nabla f(\mathbf{x}) + \sum_{i=1}^{p} \lambda_i \nabla g_i(\mathbf{x}) + \sum_{i=1}^{q} \mu_i \nabla h_i(\mathbf{x}) = \mathbf{0} \qquad (3.268b)$$

$$\nabla_{\lambda_k} L(\mathbf{x}, \boldsymbol{\lambda}, \boldsymbol{\mu}, \boldsymbol{\xi}) = g_k(\mathbf{x}) = 0, \quad 1 \leq k \leq p \qquad (3.268c)$$

$$\nabla_{\mu_k} L(\mathbf{x}, \boldsymbol{\lambda}, \boldsymbol{\mu}, \boldsymbol{\xi}) = h_k(\mathbf{x}) + \xi_k^2 = 0, \quad 1 \leq k \leq q \qquad (3.268d)$$

$$\nabla_{\xi_k} L(\mathbf{x}, \boldsymbol{\lambda}, \boldsymbol{\mu}, \boldsymbol{\xi}) = \mu_k \xi_k = 0, \quad 1 \leq k \leq q \qquad (3.268e)$$

It is important to notice that,

$$h_k(\mathbf{x}) + \xi_k^2 = 0 \Rightarrow \xi_j^2 = -h_k(\mathbf{x}) \qquad (3.269)$$

and,

$$\mu_k \xi_k = 0 \Leftrightarrow \mu_k^2 \xi_k^2 = 0 \qquad (3.270)$$

hence,

$$-\mu_k^2 h_k(\mathbf{x}) = 0 \qquad (3.271)$$

If $\mu_k \neq 0$, then

$$\mu_k h_k(\mathbf{x}) = 0 \qquad (3.272)$$

On the other hand, if $\mu_k = 0$, then $\mu_k h_k(\mathbf{x}) = 0$. Notice that, if the constraint k is active, then $\xi_k^2 = 0$ and, thus, $\mu_k \neq 0$. Otherwise, if the constraint is inactive, $\xi_k^2 > 0$ and $\mu_k = 0$. With these considerations in mind, (3.268) can be rewritten as:

$$\nabla f(\mathbf{x}) + \sum_{i=1}^{p} \lambda_i \nabla g_i(\mathbf{x}) + \sum_{i=1}^{q} \mu_i \nabla h_i(\mathbf{x}) = 0 \qquad (3.273\text{a})$$

$$g_k(\mathbf{x}) = 0, \quad 1 \leq k \leq p \qquad (3.273\text{b})$$

$$h_k(\mathbf{x}) \leq 0, \quad 1 \leq k \leq q \qquad (3.273\text{c})$$

$$\mu_k h_k(\mathbf{x}) = 0, \quad 1 \leq k \leq q \qquad (3.273\text{d})$$

Observe that the slack variables are no longer included in this new set of equations. Meanwhile, there is still an issue that it is not expressed in the above formulation. To make it clearer, suppose an inequality constraint with the form:

$$g(\mathbf{x}) \leq 0$$

The gradient vector $\nabla g(\mathbf{x})$ points to the *opposite* direction on the space region spanned by the constraint[11]. Figure 3.28 presents a distinct set of constraint functions and their respective gradient vectors along several points on the 2D space.

Within this reference frame, let's take the minimization of the function $f(x_1, x_2) = x_1^2 + x_2^2$ subject to the constraint $-x_1 + x_2 - 2 \geq 0$. This problem is graphically represented in Figure 3.28(a). Additionally, Figure 3.29 reveals the level curves associated with the objective function over the admissible region. By visual inspection, it is possible to observe that the optimum of this problem is located at point $(-1, 1)$. Moreover, the gradient direction of $f(x_1, x_2)$ is opposite to the gradient direction of $g(x_1, x_2)$. Thus, the Lagrangian multiplier is expected to be negative. In order to substantiate these assumptions, the above problem will be solved following the Lagrange multipliers technique. In order to do this, slack variables are added to the inequality constraints as shown below:

$$\begin{aligned} \min_{x_1, x_2} \quad & f(x_1, x_2) = x_1^2 + x_2^2 \\ \text{s.t.} \quad & g(x_1, x_2) = -x_1 + x_2 - 2 - \xi^2 = 0 \qquad (3.274) \\ & \xi^2 \geq 0 \end{aligned}$$

---

[11]Of course, if $g(\mathbf{x})$ is *always* less than or equal to zero, regardless of the value of $\mathbf{x}$, then the constraint existence is redundant.

leading to the following Lagrangian function:

$$L(x_1, x_2, \lambda, \phi) = \left(x_1^2 + x_2^2\right) + \lambda\left(-x_1 + x_2 - 2 - \xi^2\right) \qquad (3.275)$$

In order to find the Lagrangian critical point the following set of equations are established:

$$
\begin{aligned}
\frac{\partial}{\partial x_1} L(x_1, x_2, \lambda, \xi) &= 2x_1 - \lambda = 0 \\
\frac{\partial}{\partial x_2} L(x_1, x_2, \lambda, \xi) &= 2x_2 + \lambda = 0 \\
\frac{\partial}{\partial \lambda} L(x_1, x_2, \lambda, \xi) &= -x_1 + x_2 - 2 - \xi^2 = 0 \\
\frac{\partial}{\partial \phi} L(x_1, x_2, \lambda, \xi) &= -2\xi\lambda = 0
\end{aligned}
\qquad (3.276)
$$

which leads to:

$$
\begin{cases}
x_1 = \frac{\lambda}{2} \\
x_2 = -\frac{\lambda}{2} \\
\lambda^2\left(-\lambda - 2\right) = 0 \\
\xi^2\lambda^2 = 0
\end{cases}
\Rightarrow \lambda = 0 \vee \lambda = -2 \qquad (3.277)
$$

If $\lambda = 0$ then $x_1 = 0$, $x_2 = 0$ and $-2 - \xi^2 = 0$ which implies that $\xi^2 = -2$. Note that, in this case, the constraint $\xi^2 \geq 0$ is not validated, and thus this solution is not admissible. On the other hand, if $\lambda = -2$, $x_1 = -1$, $x_2 = 1$ and $\xi^2 = 0$. Since the constraint $\xi^2 \geq 0$ is validated, the solution (-1,1) is admissible and defines the location of a function extremum.

As predicted, both the problem solution and the Lagrange multiplier sign are validated. During the Lagrangian definition, if the term $-\lambda\left(-x_1 + x_2 - 2 - \xi^2\right)$ had been chosen instead of its positive version, the conclusion about the Lagrange multiplier would be different. More specifically, λ would have to be positive.

Rerunning the previous problem, but now assuming that the constraint has a different inequality sign,

$$
\begin{aligned}
\min \quad & f(x_1, x_2) = x_1^2 + x_2^2 \\
\text{s.t.} \quad & g(x_1, x_2) = -x_1 + x_2 - 2 \leq 0
\end{aligned}
\qquad (3.278)
$$

leads to the new graphical problem description that can be observed from Figure 3.30.

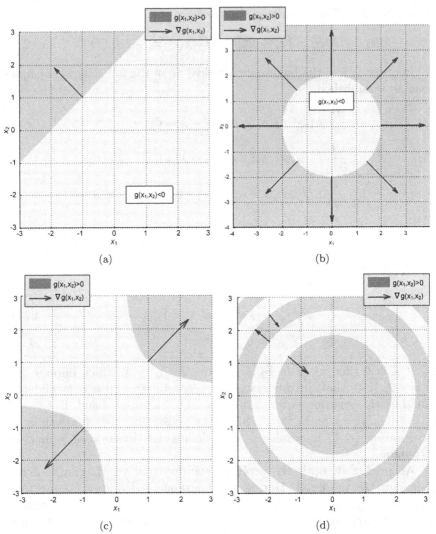

Figure 3.28 At the top left, $g(x_1, x_2) = -x_1 + x_2 - 2 \leq 0$. In (b), the function $g(x_1, x_2)$ defines a disk centered on the origin with radius 2, i.e. $g(x_1, x_2)$. Below, on the left, is the hyperbola case $x_1 \cdot x_2 - 1 \leq 0$. Figure (d) shows a nonconvex function. Specifically the cardinal sine function mathematically defined by $g(x_1, x_2) = \text{sinc}\left(0.3 \cdot \left(x_1^2 + x_2^2\right)\right) \leq 0$.

Color version at the end of the book

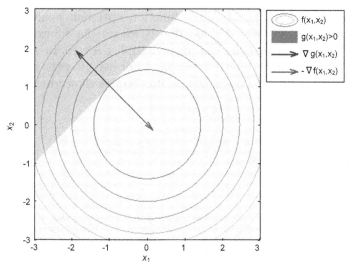

Figure 3.29 Level curves associated with the objective function $f(x_1, x_2) = x_1^2 + x_2^2$ overlapping the admissible region given by $-x_1 + x_2 - 2 \geq 0$.

Color version at the end of the book

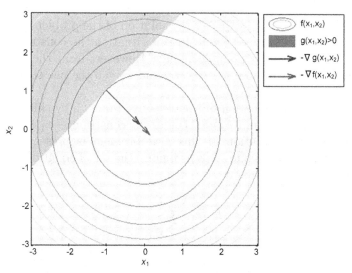

Figure 3.30 Level curves associated with the objective function $f(x_1, x_2) = x_1^2 + x_2^2$ overlapping the admissible region given by $-x_1 + x_2 - 2 \leq 0$.

Color version at the end of the book

In this situation, the objective function optimum point should be $x_1 = 0$ and $x_2 = 0$. Since this location is within the admissible region, then it is anticipated that the Lagrange multiplier is zero. In order to validate this supposition, the new Lagrangian is formed after the introduction of slack variables into the optimization problem as shown below:

$$\min_{x_1, x_2} \quad f(x_1, x_2) = x_1^2 + x_2^2$$
$$s.t. \quad g(x_1, x_2) = -x_1 + x_2 - 2 + \xi^2 = 0 \tag{3.279}$$
$$\xi^2 \geq 0$$

leading to:

$$L(x_1, x_2, \lambda, \xi) = \left(x_1^2 + x_2^2\right) + \lambda\left(-x_1 + x_2 - 2 + \xi^2\right) \tag{3.280}$$

Now, solving the following set of equations:

$$\begin{cases} x_1 = \frac{\lambda}{2} \\ x_2 = -\frac{\lambda}{2} \\ \lambda^2\left(-\lambda - 2\right) = 0 \\ \xi^2 \lambda^2 = 0 \end{cases} \tag{3.281}$$

results in $\lambda = 0 \vee \lambda = -2$. If $\lambda = 0$, then $x_1 = 0$ and $x_2 = 0$. Thus, $\xi^2 = 2$ which is positive. If $\lambda = -2$ the solution of the problem is $x_1 = -1$ and $x_2 = 1$ which leads to $\xi^2 = 0$. In this situation the constraint $\xi^2 \geq 0$ is validated for the two possible points. However, from the two above defined admissible solution, it is $x_1 = 0$ and $x_2 = 0$ the one that provides the smallest value for $f(x_1, x_2)$.

Consider the modification of the objective function by shifting its optimal point to $x_1 = -2$ and $x_2 = 2$. Under the constraint $-x_1 + x_2 - 2 \geq 0$, the graphical representation of this new problem is presented in Figure 3.31. Following the same steps as before, the set of equations obtained from the Lagrangian, leads to:

$$x_1 = \frac{\lambda - 4}{2}$$
$$x_2 = \frac{4 - \lambda}{2} \tag{3.282}$$
$$\lambda^2\left(-\lambda + 2\right) = 0 \Rightarrow \lambda = 0 \vee \lambda = 2$$

For $\lambda = 0$, $x_1 = -2$ and $x_2 = 2$. Given that $-x_1 + x_2 - 2 = \xi^2$, then $\xi^2 = 2$ which guarantees the fulfillment of the constraint $\xi^2 \geq 0$. So, this

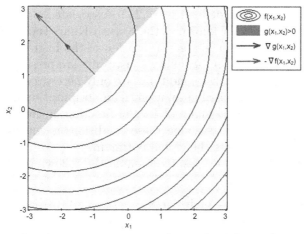

Figure 3.31 Level curves associated with the objective function $f(x_1, x_2) = (x_1 + 2)^2 + (x_2 - 2)^2$ superimposed on the permissible region given by $-x_1 + x_2 - 2 \geq 0$.

Color version at the end of the book

solution is admissible. If $\lambda = 2$, $x_1 = -1$ and $x_2 = 1$ is also an admissible solution since $\xi^2 = 0$. However, the former solution provides a lower value for the objective function.

The last situation to be considered is to solve the following quadratic problem:

$$\begin{aligned} \min \quad & f(x_1, x_2) = (x_1 + 2)^2 + (x_2 - 2)^2 \\ s.t. \quad & g(x_1, x_2) = -x_1 + x_2 - 2 \leq 0 \end{aligned} \tag{3.283}$$

In this case, the two possible values for λ are 0 and 2. Assuming $\lambda = 0$, the optimum will be at $x_1 = -2$ and $x_2 = 2$. However, for this situation $\xi^2 = -2$ which implies that this solution is not admissible. On the other hand, for $\lambda = 2$, $\xi = 0$ which means that the point located at $x_1 = -1$ and $x_2 = 1$ is admissible and is the optimum of the problem.

After analysing the previous presented examples, it is possible to conclude that, for minimization problems, the Lagrange multipliers associated to "lower or equal" type constraints must be positive or zero. In the same line of thought, for "greater or equal" constraints, the multipliers must be negative or zero. It should also be noted that, in the case of maximization type problems, the multipliers must be positive for the "greater or equal" constraints and negative for the "lower or equal" constraints.

The set of equations defined in (3.273), associated to the Lagrange multipliers signal type, are known as the Karush-Kuhn-Tucker conditions. Those conditions stand up as a generalization of the Lagrange multiplier method. In particular when inequality constraints are presented.

The Karush-Kuhn-Tucker conditions are only necessary and not sufficient. For this reason, a possible solution to an optimization problem that checks all those conditions may not be the function optimum. However, for convex type problems, checking these conditions is enough to ensure that the solution represents the global optimum.

To summarize the above results, assume the following generic minimization problem:

$$\min_{\mathbf{x}} \quad f(\mathbf{x})$$
$$s.t. \quad \mathbf{g(x) = 0}$$
$$\mathbf{h(x) \le 0} \tag{3.284}$$
$$\mathbf{q(x) \ge 0}$$

where $\mathbf{g(x)} = [g_1(\mathbf{x}) \quad \cdots \quad g_m(\mathbf{x})]^T$, $\mathbf{h(x)} = [h_1(\mathbf{x}) \quad \cdots \quad h_n(\mathbf{x})]^T$ and $\mathbf{q(x)} = [q_1(\mathbf{x}) \quad \cdots \quad q_l(\mathbf{x})]^T$ for m, n and l being the number of equality, negativity and positivity constraints, respectively. In this case, the Lagrangian is:

$$L(\mathbf{x}, \boldsymbol{\lambda}, \boldsymbol{\mu}, \boldsymbol{\alpha}) = f(\mathbf{x}) + \boldsymbol{\lambda}^T \mathbf{g(x)} + \boldsymbol{\mu}^T \mathbf{h(x)} + \boldsymbol{\alpha}^T \mathbf{q(x)} \tag{3.285}$$

where the vectors $\boldsymbol{\lambda} = [\lambda_1 \quad \cdots \quad \lambda_m]^T$, $\boldsymbol{\mu} = [\mu_1 \quad \cdots \quad \mu_n]^T$, and

$$\boldsymbol{\alpha} = [\alpha_1 \quad \cdots \quad \alpha_l]^T$$

refers to the Lagrange multipliers associated with the problem constraints. In order for a point $\mathbf{x} = \mathbf{x}^*$ to be an optimum, the following Karush-Kuhn-Tucker conditions must be validated:

Stationarity: The gradient of the objective function must be parallel to the linear combination of the gradient vectors associated to the constraints. That is,

$$\nabla f(\mathbf{x}^*) + \boldsymbol{\lambda}^T \nabla \mathbf{g(x^*)} + \boldsymbol{\mu}^T \nabla \mathbf{h(x^*)} + \boldsymbol{\alpha}^T \nabla \mathbf{q(x^*)} = 0 \tag{3.286}$$

Orthogonality: At the function extremum, the Lagrange multipliers must be orthogonal to the inequality constraints. That is,

$$\mu_i \cdot h_i(\mathbf{x})|_{\mathbf{x}=\mathbf{x}^*} = 0, \forall i$$
$$\alpha_j \cdot q_j(\mathbf{x})|_{\mathbf{x}=\mathbf{x}^*} = 0, \forall j \tag{3.287}$$

Primal Admissibility: The optimum must be an admissible solution in the original problem formulation. The same is to say that,

$$g(x^*) = 0$$
$$h(x^*) \leq 0 \qquad (3.288)$$
$$q(x^*) \geq 0$$

Dual Admissibility: The dual of an optimization problem is obtained by forming the Lagrangian using non-negative multipliers associated to the objective function constraints. The dual solution must also be admissible. That is,

$$\mu \geq 0$$
$$\alpha \leq 0 \qquad (3.289)$$

If the problem is not a minimization but a maximization problem, all Karush-Kuhn-Tucker conditions, with the exception of dual space admissibility, are identical. If a maximization problem is considered, for a solution to be admissible in the dual space, the negativity of the Lagrange multipliers must be replaced by $\mu \leq 0$ and $\alpha \geq 0$.

3.5.5.3 Linear inequality constraints

Let's return the Rosen's algorithm and consider the situation where the constraints have the following formulation:

$$Ax - b \leq 0 \qquad (3.290)$$

The Rosen's method is based on the projection of the objective function gradient over the constraints that are active at the present iteration. Consider an optimization problem expressed in canonical form by:

$$\min \quad f(x)$$
$$s.t. \quad g_i(x) \leq 0, i = 1, \cdots, m \qquad (3.291)$$

The ith constraint is called active (at the point x_l) if $g_i(x_l) = 0$. On the other hand, a passive constraint is one that, for a given solution x_l, leads to $g_i(x_l) < 0$.

Consider a point located in the admissible region. Moreover, assume that this point is not on the border of that region. If this is the case, the direction of the gradient vector does not need to be deflected by the projection matrix. This is true since it will surely point to an admissible location. Figure 3.32 illustrates this situation. In this reference frame, only η must be tuned in order to ensure that the new solution is not cast away of the admissible region.

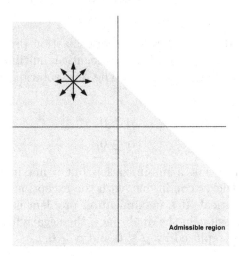

Figure 3.32 A point on the admissible region far from the border. Identification of all locations where this point can be moved without losing admissibility.

If, on the other hand, the solution is at the boundary of the admissible region, then at least one of the constraints is active. In this situation, the direction of the gradient must be bent in order to point only to the admissible region. Figure 3.33 illustrates this aspect.

It has already been seen that the projection matrix translates the gradient vector $-\nabla f(\mathbf{x}_l)$ to the intersection of all hyperplane perpendicular to $\mathbf{A}$. Thus, in the case of an optimization problem with inequality constraints, the search direction at a given iteration l, is obtained by projecting the gradient vector over the constraints that are currently active. Comparatively to the case where all constraints were of equality type, the re-estimation formulas for the inequality constraints case become:

$$\mathbf{d}_l = \frac{\left(\mathbf{N}_l^T \left(\mathbf{N}_l \mathbf{N}_l^T \right)^{-1} \mathbf{N}_l - \mathbf{I} \right) \nabla f(\mathbf{x}_l)}{\left\| \left(\mathbf{N}_l^T \left(\mathbf{N}_l \mathbf{N}_l^T \right)^{-1} \mathbf{N}_l - \mathbf{I} \right) \nabla f(\mathbf{x}_l) \right\|} \tag{3.292a}$$

$$\mathbf{x}_{l+1} = \mathbf{x}_l + \eta_l \cdot \mathbf{d}_l \tag{3.292b}$$

Figure 3.33 A point on the limit of the admissible region. Identification of all locations where this point can be moved without losing admissibility.

where **I** refers to the identity and $\mathbf{N}_l$ refers to the matrix formed by the lines of the matrix **A** which, at iteration l, refers to active constraints. For example, if the constraints i, j and k are active at iteration l then,

$$\mathbf{N}_l = \begin{bmatrix} \mathbf{A}_i \\ \mathbf{A}_j \\ \mathbf{A}_k \end{bmatrix} \tag{3.293}$$

where $\mathbf{A}_i$, $\mathbf{A}_j$ and $\mathbf{A}_k$ refer to the vectors obtained from lines i, j and k of the matrix **A**.

Notice that, unlike the case where there are only equality constraints, in this new situation the projection matrix has to be computed at all iterations and not only once as before. This is true since, at each iteration, there may be constraints that switch from active to passive state and vice-versa.

The initial admissible solution will be adjusted iteratively, according to (3.292). In theory, the optimal solution, at least locally, is considered to be found when the solution remains constant, i.e. $\mathbf{x}_{l+1} = \mathbf{x}_l$. Since $\eta_l > 0$, the only way for this to happen is if $\mathbf{d}_l = \mathbf{0}$. However, this condition can also be verified if the solution is sub-optimal. For example, if $\mathbf{d}_l = \mathbf{0}$ then,

$$\left(\mathbf{N}_l^T \left(\mathbf{N}_l \mathbf{N}_l^T \right)^{-1} \mathbf{N}_l - \mathbf{I} \right) \nabla f (\mathbf{x}_l) = \mathbf{0} \tag{3.294}$$

On the other hand, the Lagrange multipliers for this type of problems, are obtained from:

$$\boldsymbol{\lambda} = - \left(\mathbf{N}_l \mathbf{N}_l^T \right)^{-1} \mathbf{N}_l \nabla f\left(\mathbf{x}_l\right) \qquad (3.295a)$$

$$\lambda_0 = \frac{1}{2} \left\| \left(\mathbf{N}_l^T \left(\mathbf{N}_l \mathbf{N}_l^T \right)^{-1} \mathbf{N}_l - \mathbf{I} \right) \nabla f\left(\mathbf{x}_l\right) \right\| \qquad (3.295b)$$

Then, (3.294) becomes,

$$-\nabla f\left(\mathbf{x}_l\right) = \mathbf{N}_l^T \boldsymbol{\lambda} \qquad (3.296)$$

If $\mathbf{d}_l = \mathbf{0}$, then the direction of the objective function gradient will be equal to the weighted sum of the gradient of the active constraints having the Lagrange multipliers as weighting factors. If all the Lagrangian multipliers are non-negative, then the Karush-Kuhn-Tucker conditions are validated. This implies that the iterative process should be stopped since $\mathbf{x}_l$ is the optimum solution. However, if there are some negative multipliers, and if $\mathbf{d}_l = \mathbf{0}$, then $\mathbf{x}_l$ is not the optimum. In this situation, the constraints associated with these multipliers are removed from the matrix $\mathbf{N}_l$. That is, at iteration $l + 1$, the matrix $\mathbf{N}_l$ is based only on the gradient of the active constraints whose multipliers are non-negative.

In practice, not all constraints whose Lagrange multipliers are negative are discarded. As a matter of fact, only the constraint whose multiplier has the highest absolute value is removed from $\mathbf{N}_l$. If there are no active constraints, then the problem is reduced to an unconstrained optimization problem. In this case, caution must be taken when defining the value of η. It may happen that the value is too high and the solution moves to a region of space not admissible. In order to illustrate this situation consider the problem represented in Figure 3.34.

The admissible region, represented by the shading area, is bounded by three lines. Suppose that, at the present iteration, the problem solution is at $(-0.5, 1)$. Also, consider that the direction to be followed by the solution, in the next iteration, is $\mathbf{d} = \frac{\sqrt{2}}{2}[1 \quad 1]^T$.

As can be seen from Figure 3.33, the "jump" that the current solution can take along the direction $\mathbf{d}$ must be carefully chosen in order for it to remain within the admissible region. If the current iteration solution is $\mathbf{x} = [x_1 \quad x_2]^T$, then η must be such that,

$$\mathbf{A}\mathbf{x}_{l+1} - \mathbf{b} \le \mathbf{0} \qquad (3.297)$$

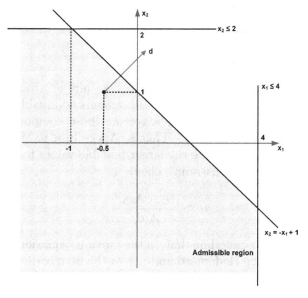

Figure 3.34 Definition of the admissible region by the intersection of three half-planes.

According to Figure 3.33, the current constraints are:

$$x_2 - 2 \leq 0$$
$$x_1 - 4 \leq 0 \tag{3.298}$$
$$x_1 + x_2 - 1 \leq 0$$

Which leads to the following representation of matrix $\mathbf{A}$ and vector $\mathbf{b}$:

$$\mathbf{A} = \begin{bmatrix} 0 & 1 \\ 1 & 0 \\ 1 & 1 \end{bmatrix} \quad \mathbf{b} = \begin{bmatrix} 2 \\ 4 \\ 1 \end{bmatrix} \tag{3.299}$$

On the other hand, it is known that the new solution is obtained from:

$$\mathbf{x}_{l+1} = \mathbf{x}_l + \eta \mathbf{d}_l \tag{3.300}$$

If this new solution is also admissible, then the inequalities at (3.297) must be validated. That is,

$$\mathbf{A}\mathbf{x}_l - \mathbf{b} + \eta \mathbf{A}\mathbf{d}_l \leq \mathbf{0} \tag{3.301}$$

If there are constraints that become active for $\mathbf{x} = \mathbf{x}_l$, then there are lines in $\mathbf{A}\mathbf{x}_l - \mathbf{b}$ that are equal to zero. If $\mathbf{j}$ is a vector with the indices of these lines, then:

$$\eta \mathbf{A_j d}_l \leq 0 \tag{3.302}$$

Notice that, if $\mathbf{A_j d}_l < 0$, then η can take any arbitrary large non-negative value that the solution $\mathbf{x}_{l+1}$ still remains admissible.

On the other hand, consider $\mathbf{i}$ a vector whose components are the indexes of the inactive constraints. That is, $\mathbf{A_i x}_l - \mathbf{b_i} < 0$. Moreover, let $\boldsymbol{\eta}$ be a vector whose elements are the larger possible values for the learning ratio associated to each constraint. Then,

$$\eta \leq \frac{(\mathbf{b_i} - \mathbf{A_i x}_l)}{\mathbf{A_i d}_l} \tag{3.303}$$

It is important to underline that, in the previous equation, the fraction operation refers to the Hadamard and not to the matrix division. For the case illustrated in Figure 3.33, the above condition leads to:

$$\eta \leq \frac{\left(\begin{bmatrix} 2 \\ 4 \\ 1 \end{bmatrix} - \begin{bmatrix} 0 & 1 \\ 1 & 0 \\ 1 & 1 \end{bmatrix} \begin{bmatrix} -0.5 \\ 1 \end{bmatrix} \right)}{\begin{bmatrix} 0 & 1 \\ 1 & 0 \\ 1 & 1 \end{bmatrix} \begin{bmatrix} 1 \\ 1 \end{bmatrix} \cdot \frac{\sqrt{2}}{2}} \approx \begin{bmatrix} 1.4 \\ 6.4 \\ 0.35 \end{bmatrix} \tag{3.304}$$

Observe that in (3.303), since $\mathbf{x}_l$ is an admissible solution, the numerator is always nonnegative. However, the denominator may, or may not, be positive depending on certain constraints. This means that the sign of a particular η_i depends on the sign of the ith element of the denominator vector. Note also that the denominator can be interpreted as the rate of change of the constraints against the respective value of η. That is,

$$\frac{\partial \left(\mathbf{b_i} - \mathbf{A_i x}_l \right)}{\partial \eta} = \mathbf{A_i d}_l \tag{3.305}$$

So, if η_i is negative, it means that the direction $\mathbf{d}_l$ is such that the solution moves away from the constraint i. Otherwise, it means that the direction will lead to a solution that approaches the constraint boundary. In this latter situation, special care is required in defining the value of η or else the solution may overshoot the admissible region boundary. The way to circumvent this situation begins by noticing that each of the components of the vector $\boldsymbol{\eta}$ denotes the limit value of η associated with each of the constraints. That is, the value that η must take so that the next solution is at the border of the respective constraint. For example, considering the results presented in (3.304), if the next solution is obtained by displacing the solution $\mathbf{x}_l$ along the direction $\frac{\sqrt{2}}{2}[1 \quad 1]^T$ using $\eta = 1.4$, then the next solution will be located exactly at the constraint boundary defined by $x_2 - 2 \leq 0$. This fact is illustrated in Figure 3.35.

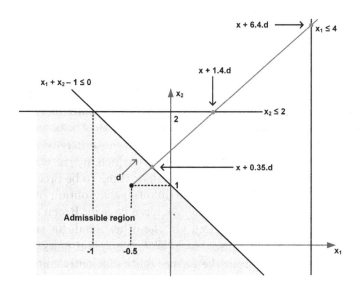

Figure 3.35 Location of the new solution based on the value of η.

If there were one (or more) negative values in the vector (3.304), this would mean that, in the next iteration, regardless of the value of the chosen η, the constraint associated with that negative value would never be violated.

As can be seen from Figure 3.34, from all the components of $\boldsymbol{\eta}$, only one of them can lead to an admissible solution. Indeed, the value for η must be selected as the smallest value of the vector $\boldsymbol{\eta}$. Therefore, the value of η, for the iteration l, must be within the following interval:

$$0 < \eta_l \leq \eta_{\max} \tag{3.306}$$

where $\eta_{\max} = \min_{\mathbf{A_i}} \left(\boldsymbol{\eta}_0^+ \right)$. The variable $\boldsymbol{\eta}_0^+$ refers to the vector composed by only the non-negative values of $\boldsymbol{\eta}$. The subscript $\mathbf{A_i}$ is used to emphasize that the maximum value for η_l, is only determined on the basis of the inactive constraints. That is, all the constraints that are not contemplated in the matrix $\mathbf{N}_i$.

If, at a given iteration, there are no inactive constraints, the η can be as large as desired. However, as was seen, values that are too high for the learning rate can lead to instability. In order to avoid oscillations, and increase convergence of the algorithm, the value of η_l is determined by solving the following optimization problem:

$$\min_{\eta_l} \quad f\left(\mathbf{x}_l + \eta_l \mathbf{d}_l\right)$$
$$s.t. \quad 0 < \eta_l \leq \eta_{\max} \tag{3.307}$$

Observe that both $\mathbf{x}_l$ and $\mathbf{d}_l$ are known and constant. Hence, this single variable optimization problem can be solved quite efficiently using relatively simple algorithms such as the Newton or the "bisection" method.

The bisection method is very appealing because it only requires the knowledge of the objective function derivative which, by the way, has to be computed to run the gradient projection algorithm. To be precise, it is not quite the same derivative but it is possible to easily obtain one from the other using the derivative chain rule. As already seen, Rosen's algorithm, requires the computation of $\nabla f(\mathbf{x})$. On the other hand, for the bisection method, it is necessary to know $\frac{\partial f(\mathbf{x}_k + \eta \mathbf{d}_k)}{\partial \eta}$. The good news is that it is possible to get the latter from the former quite efficiently since:

$$\frac{\partial f\left(\mathbf{x}\right)}{\partial \eta} = \left(\frac{\partial f\left(\mathbf{x}\right)}{\partial \mathbf{x}}\right)^T \frac{\partial \mathbf{x}}{\partial \eta} \tag{3.308}$$

and taking into consideration that $\mathbf{x}_l = \mathbf{x}_{l-1} + \eta_l \mathbf{d}_l$ then,

$$\frac{\partial \mathbf{x}_l}{\partial \eta_l} = \mathbf{d}_l \tag{3.309}$$

leading, finally, to:

$$\frac{\partial f\left(\mathbf{x}_k\right)}{\partial \eta_l} = \nabla^T f(\mathbf{x}_l)\mathbf{d}_l \tag{3.310}$$

Seemingly to the Rosen's algorithm, the bisection method is an iterative procedure and its execution steps are defined through the pseudo-code presented in Algorithm 8.

Algorithm 8 Bisection algorithm applied to the problem (3.307).

It is assumed that η must be found somewhere between η_{min} and η_{max}

1   Determine $\hat{\eta} = \frac{\eta_{max} + \eta_{min}}{2}$.

2   Calculate $\nabla f(\mathbf{x}_l + \hat{\eta} \mathbf{d}_l)$.

3   Obtain $s = \nabla^T f(\mathbf{x}_l + \hat{\eta} \mathbf{d}_l) \mathbf{d}_k$.

4   If $s < 0$ then $\eta_{min} = \hat{\eta}$.
     If $s > 0$ then $\eta_{max} = \hat{\eta}$.

5   If $s = 0$ end, else return to step (2).

3.5.5.4 Putting it all together...

During this section, all the pieces derived in the previous three sections will be assembled to produce the Rosen algorithm. As already referred, this is a gradient based search method which can be applied to constrained optimization problems. This algorithm can be used considering any linear or non-linear objective function as long as it is differentiable. However, the constraints are assumed to be linear. During its work, Rosen has proposed a method to deal with the case where the constraints are non-linear. This method involves the linearization of the non-linear constraint equations around the current solution $\mathbf{x}_l$, by means of the Taylor series expansion, as well as calculating a correction factor to keep the solution within the admissible region. However, in this book, this situation is not considered since the constraints associated with the training of a hidden Markov model are linear. Integration of nonlinear constraints within the Rosen method can be found at [5], [29] and [24].

In short, the Rosen algorithm can be applied to the class of optimization problem that can be described using the following canonical form:

$$\min_{\mathbf{x}} \quad f(\mathbf{x})$$
$$s.t. \quad \mathbf{Ax} - \mathbf{b} \leq \mathbf{0} \tag{3.311}$$
$$\mathbf{Gx} - \mathbf{h} = \mathbf{0}$$

Notice that, during the search iterative process, some of the constraints of $\mathbf{Ax} - \mathbf{b} \leq \mathbf{0}$ may become active. Thus, it is assumed that the matrix $\mathbf{A}$ can be decomposed into:

$$\mathbf{A} = \begin{bmatrix} \mathbf{A}_a \\ \mathbf{A}_i \end{bmatrix} \tag{3.312}$$

where $\mathbf{A}_a$ refers to the lines of $\mathbf{A}$ associated with *active* constraints and $\mathbf{A}_i$ refers to *inactive* constraints. That is:

$$\begin{aligned}\mathbf{A}_a\mathbf{x} - \mathbf{b} &= 0 \\ \mathbf{A}_i\mathbf{x} - \mathbf{b} &< 0\end{aligned} \tag{3.313}$$

By taking into consideration all the things that have been said in the previous sections, the Rosen algorithm can be finally put together. The sequence of steps to be performed by the Rosen algorithm, in order to obtain the solution of the optimization problem (3.311), is presented in Algorithm 9.

Algorithm 9 Rosen algorithm.

---

1 Iteration l equal to 0.
2 Establish an acceptable initial solution $\mathbf{x}_l$.
3 Determine active constraints and form $\mathbf{N}_l$.
   3.1 If there are **no** active constraints then $\mathbf{N}_l = [\ \ ]$ and $\mathbf{d}_l = -\nabla f(\mathbf{x}_l)$.
      3.1.1 If $\mathbf{d}_l \neq \mathbf{0}$ then $\mathbf{x}_l$ is optimum and the algorithm ENDS.
      3.1.2 If $\mathbf{d}_l = \mathbf{0}$ then the algorithm continues in step **4**.
   3.2 If **there are** active constraints obtains $\mathbf{N}_l = \begin{bmatrix} \mathbf{A}_a \\ \mathbf{G} \end{bmatrix}$ and calculates:

$$\mathbf{P}_l = \mathbf{N}_l^T \left(\mathbf{N}_l\mathbf{N}_l^T\right)^{-1} \mathbf{N}_l - \mathbf{I}$$
$$\mathbf{d}_l = \mathbf{P}_l\nabla f(\mathbf{x}_l)$$

      3.2.1 If $\mathbf{d}_l = \mathbf{0}$ then the Lagrange multipliers are determined from:

$$\boldsymbol{\lambda} = - \left(\mathbf{N}_l\mathbf{N}_l^T\right)^{-1} \mathbf{N}_l\nabla f(\mathbf{x}_l)$$

where $\boldsymbol{\lambda}$ is a vector column whose values are the Lagrange multipliers associated with each of the active constraints.
         3.2.1.1 If $\boldsymbol{\lambda} \geq 0$ then $\mathbf{x}_l$ validates the Karush-Kuhn-Tucker conditions. The algorithm ENDS with $\mathbf{x}_l$ as optimum.
         3.2.1.2 If there exists any $\lambda_i < 0$ for $i = 1, \cdots, \dim(\mathbf{A}_a)$ then the restriction in $\mathbf{A}_a$ whose Lagrange multiplier be the most negative, it is eliminated from $\mathbf{N}_l$. Returns to step **3.1**.
      3.2.2 If $\mathbf{d}_l \neq \mathbf{0}$ then the algorithm continues in step **4**.
4 Normalize $\mathbf{d}_l$:
$$\mathbf{d}_l \leftarrow \frac{\mathbf{d}_l}{||\mathbf{d}_l||}$$
5 If there are inactive restrictions then:
   Using only inactive constraints $\mathbf{A}_i$ determine η_{max} from:
$$\eta \leq \frac{(\mathbf{b}-\mathbf{A}_i\mathbf{x}_l)}{\mathbf{A}_i\mathbf{d}_l} \text{ for all } i \text{ such that } \mathbf{A}_i\mathbf{d}_l > 0$$
$$\eta_{max} = \min \eta$$
   Using the bisection method solve:
$$\min_{\eta_l} \ f(\eta_l)$$
$$s.t. \quad 0 < \eta_l \leq \eta_{max}$$
   Otherwise establish η_{max} ad-hoc and solve, by the bisection method:
$$\min_{\eta_l} \ f(\eta_l)$$
$$s.t. \quad 0 < \eta_l \leq \eta_{max}$$
7 Create a new solution from the map:
$$\mathbf{x}_{l+1} = \mathbf{x}_l + \eta_l \cdot \mathbf{d}_l$$
8 Increase l and return to step **3**.

The above referred algorithm, encoded for MATLAB®, can be found in Listing 3.23. This function makes calls to other three MATLAB® functions which are presented in Listings 3.24, 3.25 and 3.26.

The MATLAB® function related to the Rosen algorithm can be tested by running the MATLAB® script presented in Listing 3.27. This script assumes that the matrix **A** and the vector **b** are given by:

$$\mathbf{A} = \begin{bmatrix} 1 & 1 \\ 1 & 5 \\ -1 & 0 \\ 0 & -1 \end{bmatrix}, \quad \mathbf{b} = \begin{bmatrix} 2 \\ 5 \\ 0 \\ 0 \end{bmatrix} \tag{3.314}$$

and the objective function is:

$$f(x_1, x_2) = 2x_1^2 + 2x_2^2 - 2x_1 \cdot x_2 - 4x_1 - 6x_2 \tag{3.315}$$

which is described in an external MATLAB® function presented in Listing 3.28.

Listing 3.23 Rosen algorithm MATLAB® function.

```
 1  function x=rosen(gradF,x,A,b,C,d,maxiter)
 2  % Rosen algorithm for solving:
 3  %    min f(X)
 4  % s.t.
 5  %       AX-b<=0
 6  %       CX-d =0
 7
 8  % Initialization phase...
 9  [Aa,Ai,bi,Nk]=MatGradRest(A,b,C,x);
10
11  % Gradient calculation of f(x)
12  df=feval(gradF,x);
13  for itera=1:maxiter
14      disp(['------------------------Iteration: ' num2str(itera)])
15      if isempty(Nk)
16          dk=-df;
17          if sum(abs(dk))==0, %dk=0
18              % The optimum was found
19              break;
20          else
21              x=update_sol(gradF,x,Ai,b,dk);
22              [Aa,Ai,bi,Nk]=MatGradRest(A,b,C,x);
23          end
24      else
25          % Determine the projection matrix...
26          %P=Nk'\(Nk*Nk')*Nk;
27          P=Nk'*inv(Nk*Nk')*Nk;
28          P=P-eye(size(P));
29          % Determine search Direction
30          dk=P*df;
31          % checks whether critical point...
```

```
32        if sum(abs(dk))==0, % dk=0
33            % Calculate the Lagrange multipliers
34            lambda=-inv(Nk*Nk')*Nk*df;
35            % Checks the signal of the active restrictions associated
36            % with inequalities
37            if ~isempty(Aa) % If there are active inequality
                  constraints
38                [nl,nc]=size(Aa); % nl indicates the number of
39                                  % active constraints
40                neglambda=find(lambda(1:nl)<0);
41                if ~isempty(neglambda) % there are negative multipliers
                      ,
42                    [val,ind]=min(lambda);
43                    Nk(ind,:)=[]; % Remove the constraint associated
                          with
44                                  % the most negative multiplier...
45                    Aa(ind,:)=[];
46                else % There are no negative multipliers.
47                    % Conditions of KKT are met...
48                    break;
49                end
50            end
51        else % dk~=0
52            x=update_sol(gradF,x,Ai,bi,dk);
53            [Aa,Ai,bi,Nk]=MatGradRest(A,b,C,x);
54        end
55    end
56    df=feval(gradF,x);
57 end
```

Listing 3.24 Bisection method to be used within the Rosen algorithm.

```
1 function eta=bisectrosen(gradfunc,x,dk,limI,limS,maxiter)
2 % Bisection method
3
4 for k=1:maxiter,
5     eta=.5*(limI+limS);
6     x_k=x+eta*dk;
7     r=feval(gradfunc,x_k);
8     s=r'*dk;
9     if s==0,
10        break;
11    elseif s>0,
12        limS=eta;
13    else
14        limI=eta;
15    end
16 end
```

Listing 3.25 Compute the matrix with inactive constraints.

```
1 function [Aa,Ai,bi,Nk]=MatGradRest(A,b,C,x)
2 % Definition of the matrix with inactive constraints
3 if ~isempty(A)
4     Ai=A(find(A*x-b~=0),:);
```

```
 5      bi=b(find(A*x-b~=0),:);
 6      % Definition of the matrix with active constraints
 7      Aa=A(find(A*x-b==0),:);
 8  else
 9      Aa=[];
10      Ai=[];
11      bi=[];
12  end
13  % Construction of the matrix Nk
14  Nk=[Aa;C];
```

Listing 3.26 Function to update the new solution value.

```
 1  function x=update_sol(gradF,x,Ai,bi,dk)
 2  % eta value by default...
 3  dk=dk/norm(dk);
 4  % Determine eta max
 5  if ~isempty(Ai)
 6      eta=(bi-Ai*x)./(Ai*dk);
 7      etamax=min(eta(eta>0));
 8      % It uses the bisection method to determine the optimal eta
 9      % min f(eta)
10      % s.t. 0<=eta<=etamax
11      eta=bisectrosen(gradF,x,dk,0,etamax(1),100);
12  else
13      eta=bisectrosen(gradF,x,dk,0,1,100);
14  end
15  % Calculates new solution
16  x=x+eta*dk;
```

Listing 3.27 MATLAB® script to test the Rosen algorithm.

```
 1  % Used to test the Rosen algorithm
 2  A=[1 1;1 5;-1 0;0 -1];
 3  b=[2;5;0;0];
 4  C=[];
 5  d=[];
 6  gradF='rosen_test_func';
 7  x0=[-2;-2];
 8  MaxIter=100;
 9  x=rosen(gradF,x0,A,b,C,d,MaxIter);
10  % ----------------------------------------------------PLOTS
11  [x1,x2]=meshgrid(-1:.01:2);
12  y=double((x1+x2<=2)& (x1+5*x2<=5) & x1>=0 & x2>=0);
13  pcolor(x1,x2,y);  % Admissible region
14  shading interp;
15  xlabel('x1');
16  ylabel('x2');
17  hold on;
18  z=2*x1.^2+2*x2.^2-2*x1.*x2-4*x1-6*x2; % Objective function
19  surfl(x1,x2,z);
20  shading interp;
21  % Solution
22  z=2*x(1).^2+2*x(2).^2-2*x(1)*x(2)-4*x(1)-6*x(2);
23  plot3(x(1),x(2),0,'kx')
```

Listing 3.28 The *testbench* function.

```
1 function gf=rosen_test_func(x)
2
3 gf=[4*x(1)-2*x(2)-4;4*x(2)-2*x(1)-6];
```

The solution, after running the Rosen algorithm along 100 iterations, was $x_1 = 1.1290$ and $x_2 = 0.7742$. The graphical location of this solution is presented in Figure 3.36.

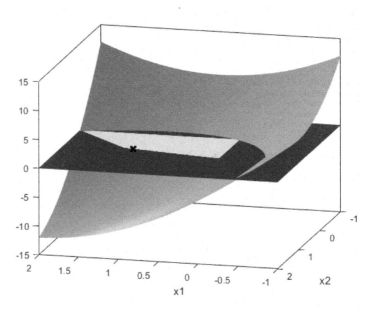

Figure 3.36 Solution of a minimization problem using the Rosen algorithm.

3.5.5.5 Rosen's method applied to hidden Markov models

Finally, it is time to explain how to adapt the Rosen algorithm to be used as a parameter estimation procedure for hidden Markov models.

As already explained, the training strategy of a hidden Markov model involves the maximization of the likelihood function. However, the parameters space is bounded. In particular, it is known that the parameter matrices, previously designated by $\mathbf{A}$, $\mathbf{B}$ and $\mathbf{c}$, must obey to the stochasticity constraint. That is, their elements must be defined within the interval $[0, 1]$ and the sum, along the lines, of each of these matrices must be equal to one. Bearing this in mind, the hidden Markov model training problem can be defined as an optimization problem with the following structure:

$$\max_{\mathbf{A},\mathbf{B},\mathbf{c}} \quad \mathcal{L}\left(\mathbf{A},\mathbf{B},\mathbf{c}\right)$$

$$\mathbf{A} \geq \mathbf{0}$$
$$\mathbf{B} \geq \mathbf{0}$$
$$\mathbf{c} \geq \mathbf{0} \tag{3.316}$$
$$\mathbf{A} \cdot \mathbf{1}^T = \mathbf{1}^T$$
$$\mathbf{B} \cdot \mathbf{1}^T = \mathbf{1}^T$$
$$\mathbf{c} \cdot \mathbf{1}^T = 1$$

To use the Rosen algorithm, in the hidden Markov model context, a different parameterization is considered. This is done by assuming three vectors $\mathbf{x_A}$, $\mathbf{x_B}$ and $\mathbf{x_c}$ with the following structure:

$$\mathbf{x_A} = \begin{bmatrix} a_{11} & \cdots & a_{1m} & \cdots & a_{m1} & \cdots & a_{mm} \end{bmatrix}^T \tag{3.317}$$

$$\mathbf{x_B} = \begin{bmatrix} b_{11} & \cdots & b_{1n} & \cdots & b_{m1} & \cdots & b_{mn} \end{bmatrix}^T \tag{3.318}$$

$$\mathbf{x_c} = \begin{bmatrix} c_1 & \cdots & c_m \end{bmatrix}^T \tag{3.319}$$

where m is the number of hidden states and n the number of observable states. Moreover, the coefficients a_{ij}, b_{ik} and c_i for $(i,j) = 1, \cdots, m$ and $k = 1, \cdots, n$ are the hidden Markov model transition probabilities as defined in Section 3.2. The new optimization problem is then defined as:

$$\min_{\mathbf{x_A},\mathbf{x_B},\mathbf{x_I}} \quad -\mathcal{L}\left(\mathbf{x_A},\mathbf{x_B},\mathbf{x_I}\right)$$

$$s.t. \quad \mathbf{Q}\mathbf{x_A} - \mathbf{1} = \mathbf{0}$$
$$\mathbf{R}\mathbf{x_B} - \mathbf{1} = \mathbf{0}$$
$$\mathbf{h}^T \mathbf{x_I} - 1 = 0 \tag{3.320}$$
$$-\mathbf{x_A} \leq \mathbf{0}$$
$$-\mathbf{x_B} \leq \mathbf{0}$$
$$-\mathbf{x_I} \leq \mathbf{0}$$

where the matrices $\mathbf{Q}$ and $\mathbf{R}$ are computed as:

$$\mathbf{Q} = \mathbf{D} \otimes \mathbf{v_m}$$
$$\mathbf{R} = \mathbf{D} \otimes \mathbf{v_n} \tag{3.321}$$

for $\mathbf{v}_m$ and $\mathbf{v}_n$ as unity vectors. The former with dimension $1 \times m$ and the latter $1 \times n$. $\mathbf{D}$ is an $m \times m$ identity matrix and $\otimes$ refers to the Kronecker product. According to this new formulation, the Rosen algorithm can then be used for parameter estimation of a hidden Markov model. The procedure is summarized in Algorithm 10 and, as usual, the MATLAB® implementation of this algorithm can be found in Listing 3.29. Notice that this main function execution requires the sub-functions presented in Listings 3.10, 3.25, 3.30, 3.31 and 3.32. The hidden Markov model parameter estimation based on the Rosen algorithm can be tested by running the script presented in Listing 3.33.

Algorithm 10 Rosen algorithm applied to hidden Markov model parameter estimation.

1   Initialize $\mathbf{A}$, $\mathbf{B}$ and $\mathbf{c}$.
2   Calculate $\bar{\Omega}_a$, $\bar{\Omega}_b$ and $\frac{\partial \mathcal{L}}{\partial \mathbf{c}}$.
3   Parameter update:
    3.1 Get new values of $\mathbf{A}$ using the Rosen algorithm.
    3.2 Get new values of $\mathbf{B}$ using the Rosen algorithm.
    3.2 Get new values of $\mathbf{c}$ using the Rosen algorithm.
4   Return to (2).

Listing 3.29 Train a hidden Markov model with the Rosen algorithm.

```
1  function [A,B,c,Fit]=hmm_rosen(m,n,O,MaxIter)
2
3  % Parameters initialization...
4  Fit=zeros(1,MaxIter);
5
6  % Initialization (random) of the matrices A, B and c
7  A=rand(m,m);A=A./(A*ones(m,1)*ones(1,m));
8  B=rand(m,n);B=B./(B*ones(n,1)*ones(1,n));
9  c=rand(1,m);c=c./(c*ones(m,1));
10
11  % creates x_A, x_N and x_c
12  x_A=reshape(A',m*m,1);
13  x_B=reshape(B',m*n,1);
14  x_c=c.';
15
16  % generates the constraint matrices: QA,QB and vector QI
17  QA=kron(eye(m),ones(1,m));
18  QB=kron(eye(m),ones(1,n));
19  QI=ones(1,m);
20
21  % generates the constraint matrices: bA,bB and vector bc
22  bA=ones(m*m,1);
23  bB=ones(m*n,1);
24  bc=1;
25
```

```
26  % generates the constraint matrices: RA, RB and vector Rc
27  RA=-eye(m*m);
28  RB=-eye(m*n);
29  Rc=-eye(m);
30
31  % generates the constraint matrices: cA, cB and vector cc
32  cA=zeros(m*m,1);
33  cB=zeros(m*n,1);
34  cc=zeros(m,1);
35
36  % Initialization Phase: A
37  [A_Aa,A_Ai,A_bi,A_Nk]=MatGradRest(RA,cA,QA,x_A);
38
39  % Initialization Phase: B
40  [B_Aa,B_Ai,B_bi,B_Nk]=MatGradRest(RB,cB,QB,x_B);
41
42  % Initialization Phase: c
43  [c_Aa,c_Ai,c_bi,c_Nk]=MatGradRest(Rc,cc,QI,x_c);
44
45  % >>>>>>>>>>>>>>>>>>>>>>>>>>>>>>>>>>>>>>>>>>>>>>>>>>>>>>>>>>>>>>>
46  for j=1:MaxIter,
47      % L-gradient calculation
48      [~,dLdA]=gradientLA_norm(A,B,O,c);
49      [~,dLdB]=gradientLB_norm(A,B,O,c);
50      [dLdc]=gradientLc(A,B,O,c);
51      % transform matrix into vector...(careful with the signal...)
52      row_dLdA=-reshape(dLdA',m*m,1);
53      row_dLdB=-reshape(dLdB',m*n,1);
54      row_dLdc=-dLdc.';
55      % HMM parameters
56      Theta{1}=A; Theta{2}=B;Theta{3}=c;
57      % Rosen algorithm...
58      [A_Aa2,A_Ai2,A_bi2,A_Nk2,x_A2]=...
59        one_shot_rosen('gradientLA_norm',Theta,O,RA,cA,QA,bA,A_Aa,A_Ai,
                A_bi,A_Nk,x_A,row_dLdA,1);
60      [B_Aa2,B_Ai2,B_bi2,B_Nk2,x_B2]=...
61        one_shot_rosen('gradientLB_norm',Theta,O,RB,cB,QB,bB,B_Aa,B_Ai,
                B_bi,B_Nk,x_B,row_dLdB,2);
62      [c_Aa2,c_Ai2,c_bi2,c_Nk2,x_c2]=...
63        one_shot_rosen('gradientLc',Theta,O,Rc,cc,QI,bc,c_Aa,c_Ai,c_bi,
                c_Nk,x_c,row_dLdc,3);
64      % Update Matrices...
65      A_Aa=A_Aa2;A_Ai=A_Ai2;A_bi=A_bi2;A_Nk=A_Nk2;x_A=x_A2;
66      B_Aa=B_Aa2;B_Ai=B_Ai2;B_bi=B_bi2;B_Nk=B_Nk2;x_B=x_B2;
67      c_Aa=c_Aa2;c_Ai=c_Ai2;c_bi=c_bi2;c_Nk=c_Nk2;x_c=x_c2;
68      % Rebuild A, B and c...
69      A=reshape(x_A,m,m).';
70      B=reshape(x_B,n,m).';
71      c=x_c.';
72      % Determines the performance of new parameters...
73      [~,P]=forward_algorithm_norm(A,B,O,c);
74      disp(['Iteration -- ' num2str(j) ' FIT -- ' num2str(P)]);
75      Fit(j)=P;
76      % <<<<<<<<<<<<<<<<<<<<<<<<<<<<<<<<<<<<<<<<<<<<<<Repetition...
77  end
```

Listing 3.30 Rosen algorithm for hidden Markov model training.

```
1  function [Aa,Ai,bi,Nk,x]=one_shot_rosen(gradfunc,Theta,O,A,b,C,d,Aa,Ai,
       bi,Nk,x,df,var)
2
3  if isempty(Nk)
4      dk=-df;
5      if sum(abs(dk))==0, %dk=0
6          % The optimum was found
7          return;
8      else
9          x=hmmupdate_sol(gradfunc,Theta,O,x,Ai,bi,dk,var);
10         [Aa,Ai,bi,Nk]=MatGradRest(A,b,C,x);
11     end
12 else
13     % Determine the projection matrix...
14     P=Nk'*((Nk*Nk')\Nk);
15     P=P-eye(size(P));
16     % Determine search direction
17     dk=P*df;
18     % checks whether critical point...
19     if sum(abs(dk))==0, % dk=0
20         % Calculate the Lagrange multipliers
21         lambda=-inv(Nk*Nk')*Nk*df;
22         % Checks the signal of the active restrictions associated with
               inequalities
23         if ~isempty(Aa) % If there are active inequality constraints
24             [nl,~]=size(Aa); % nl indicates the number of active
                   restrictions
25             neglambda=find(lambda(1:nl)<0);
26             if ~isempty(neglambda) % there are negative multipliers...
27                 [~,ind]=min(lambda);
28                 Nk(ind,:)=[]; % remove the constraint associated with
                       the most negative multiplier
29                 Aa(ind,:)=[];
30             else % there are no negative multipliers. KKT conditions
                   are met...
31                 return;
32             end
33         end
34     else % dk~=0
35         x=hmmupdate_sol(gradfunc,Theta,O,x,Ai,bi,dk,var);
36         [Aa,Ai,bi,Nk]=MatGradRest(A,b,C,x);
37     end
38 end
```

Listing 3.31 Update the hidden Markov model solution.

```
1  function x=hmmupdate_sol(gradfunc,Theta,O,x,Ai,bi,dk,var)
2  dk=dk/norm(dk);
3  % Determine eta max
4  if ~isempty(Ai)
5      den=Ai*dk;
6      inx=find(den>0);
7      if ~isempty(inx)
8          num=bi(inx,:)-Ai(inx,:)*x;
9          den=den(inx,:);
10         eta=num./den;
11         inx=find(eta<0);eta(inx)=[];
```

```
12        etamax=min(eta);
13    else
14        etamax=1;
15    end
16    % It uses the bisection method to determine the optimal for eta:
17    eta=hmmbissectrosen(gradfunc,Theta,O,x,dk,0,etamax,100,var);
18 else
19    eta=hmmbissectrosen(gradfunc,Theta,O,x,dk,0,1,100,var);
20 end
21 % Calculates new solution
22 x=x+eta*dk;
```

Listing 3.32 Bisection method applied to Rosen's hidden Markov model train method.

```
1 function eta=hmmbissectrosen(gradfunc,Theta,O,x,dk,limI,limS,maxiter,
       var)
2 [m,n]=size(Theta{var});
3 for k=1:maxiter,
4     eta=.5*(limI+limS);
5     x_k=x+eta*dk;
6     Theta{var}=reshape(x_k,n,m).';
7     [~,dLdV]=feval(gradfunc,Theta{1},Theta{2},O,Theta{3});
8     row_dLdV=-reshape(dLdV',m*n,1);
9     s=row_dLdV'*dk;
10    if s==0,
11        break;
12    elseif s>0,
13        limS=eta;
14    else
15        limI=eta;
16    end
17 end
```

Listing 3.33 MATLAB® script to test the Rosen's hidden Markov model training procedure.

```
1 clear all;
2 clc;
3 m=2;
4 n=3;
5 O=generate_data();
6 MaxIter=100;
7 [A,B,c,Fit]=hmm_rosen(m,n,O,MaxIter);
```

3.6 ARCHITECTURES FOR MARKOV MODELS

When a hidden Markov model is considered for a particular problem, several aspects must be equated besides which parameter estimation method will be used. For example, the specification of its structure is a fundamental question to be answered. A hidden Markov model structure must be approached from two distinct perspectives: in one hand, the definition of how many hidden and observable states will be available at each layer and, on the other, how those states are interconnected. Up to now, it has been assumed that the hidden Markov model is ergodic. That is, the hidden states are fully connected which means that a transition is possible between any pair of nodes. However, this is not the only possible approach. For example, in speech processing applications, a different structure is often used where there is only one input and one output for each hidden state. This strategy aims to define only a sequential way for the model evolution and, simultaneously, to reduce the parameter space dimensionality.

The way a given hidden Markov model should be connected is unknown and empirically established after a large number of experiments. Sometime it can be tempting to start with a fully interconnected hidden Markov model and wait for the training algorithm to establish the importance of the transitions between states. However, this strategy rarely works, even with a sufficiently large dataset, due to the fact that a large number of local maxima is generated by the likelihood function.

In computational terms, disabling the transition from hidden state i to hidden state j implies zeroing the element a_{ij} of the transition probabilities matrix $\mathbf{A}$. Note also that the Baum-Welch estimation process does not change a value that has been initialized with zero. For this reason, the training strategy of an ergodic chain is not different from that of, for example, a left to right model. Hence, the training algorithms described along this chapter can be applied to any hidden Markov model structure.

3.7 SUMMARY

Hidden Markov models play an important role in the actual dynamic systems modelling scenario. Within this computational paradigm, each observation is generated, through a set of hidden states, with a certain probability. Those probabilities are described in the form of two probability transition matrices $\mathbf{A}$ and $\mathbf{B}$ and a priors vector $\mathbf{c}$. Their values are usually unknown, and must be estimated from a set of observations taken from the system to model. Those parameters are obtained through the solution of a constrained optimization method whose objective functions are the observations likelihood.

In general, obtaining the likelihood of a hidden Markov model is computationally expensive and requires efficient ways for data processing. It is in this context that both the forward and backward algorithms have appeared in this chapter. The former was used for computing $P(\Lambda_k \wedge q_k = s_i)$, denoted by $\alpha_k(i)$, and the latter for obtaining $P(\Lambda_{k+1 \to N} | q_{k=s_i})$, expressed in a more compact form as $\beta_k(i)$. Increased intuition regarding the implementation of both algorithms was provided by presenting their computer formulation in the form of MATLAB® functions.

This chapter has also presented the celebrated Viterbi algorithm. The Viterbi algorithm is widely used in communication systems to decode a data sequence and can be found in several commercial applications such as hard disk drives [9]. In the hidden Markov model framework, the Viterbi algorithm can be used to derive the most probable hidden states path from a given set of observations. A MATLAB® version of the Viterbi algorithm was also provided.

Another fundamental issue, when modelling systems with hidden Markov models, is how their parameters can be estimated. Along this chapter, three distinct techniques have been addressed: the Baum-Welch method, the Viterbi training algorithm and the Rosen procedure. It is important to understand that the parameters of a hidden Markov model cannot be obtained from a closed form mathematical expression. Hence, all the three distinct algorithms presented are iterative methods that start from a guessed admissible solution, and evolve it in such a way as to increase the model likelihood.

Many sections of this chapter rely on mathematical expressions derived informally in order to alleviate the mathematical burden while keeping them more implementation oriented. For this reason, all the algorithms were accompanied by their MATLAB® implementation. Note that the set of parameter estimation techniques, presented in this chapter, is not by any means exhaustive. For example, the use of evolutionary based techniques has not been addressed.

4 Continuous Hidden Markov Models

"The sciences do not try to explain, they hardly even try to interpret, they mainly make models. By a model is meant a mathematical construct which, with the addition of certain verbal interpretations, describes observed phenomena. The justification of such a mathematical construct is solely and precisely that it is expected to work."

–John Von Neumann

Hidden Markov models are specially tailored for situations where the modeling process has a finite discourse alphabet. That is, the number of different observable situations, at the process output, is limited. However, not all processes behave like this since, frequently, their output can change continuously within an arbitrary range. For those cases, it is possible to establish an alternative family of hidden Markov models where the observations probability distribution function is replaced by a probability density function. The formal description of this hidden Markov model extension will be addressed in this chapter. As previously, all the relevant algorithms will be mathematically described and subsequently represented using MATLAB® coded functions.

This Chapter will begin by establishing the Gaussian mixture as a common strategy to approximate arbitrary continuous distribution functions. Then, the integration between these combined probability density functions and an ordinary Markov chain will be made. This will lead to the definition of a continuous observations hidden Markov model.

In this chapter, the forward, backward and Viterbi algorithms will also be extended to handle continuous hidden Markov models. Finally, the Baum-Welch parameters estimation method will be presented to handle this class of models.

4.1 INTRODUCTION

A typical hidden Markov model has a structure that comprises two sets of states, distributed over two different layers, denoted by *hidden* and *observable*. The former describes the internal system behaviour and the latter the system output universe of discourse. The universe of discourse is the set of values that a given system output can exhibit. In this common model description, each possible system output is modelled as an observable state. This discrete nature of hidden Markov models is incompatible with certain types of systems such as the ones that exhibit continuous outputs. Sure this limitation can be mitigated by resorting to quantization of the discourse. However, this solution raises some issues as discussed ahead. For example, consider the signal represented in Figure 4.1 and its four level quantized version.

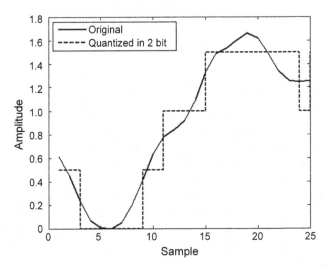

Figure 4.1 Quantization of a continuous-time signal using a 2 *bit* resolution.

In this situation, an uniform quantization was used. That is, it is assumed that the signal amplitude is bounded between the values 0 and 2. Hence, the assignments to each of the 4 possible observable states are as follows: $r_1 = 0$, $r_2 = 0.5$, $r_3 = 1$ and $r_4 = 1.5$. By comparing the original signal and its 2 bits quantized version, it is possible to observe that the amplitude discretization process will lead to a severe distortion. The distortion caused by this technique can be even more severe if the signal has a very low standard deviation as with the example shown in Figure 4.2.

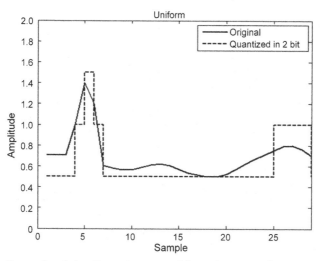

Figure 4.2 Example of the distortion caused by using a too low quantization resolution. The information carried by the signal between amplitudes 0.5 and 0.8 is lost.

In such a situation, uniform quantization is not a good choice since much of the signal information is lost. In these cases, non-uniform quantization should be used in order to overcome the distortion caused by the amplitude discretization effect. One of such approaches can be, for example, through vector quantization or resorting to other data partition techniques. To illustrate this approach, Figure 4.3 presents the same signal as the one of Figure 4.2, but now using different quantization intervals which were obtained by running a "competitive" algorithm. It can be seen that now the quantization version is able to describe, with lower distortion, the behaviour of the original signal.

Even if it is possible to quantize an analog signal, this operation almost always leads to degradation and information lost from the original signal. In practice, to minimize the quantization noise, the number of observable states must be large. As an example, CD quality audio uses 16 bits and DVD audio can push this envelope even higher to 24 bits.

However, increasing the quantization resolution brings up another problem when dealing with hidden Markov models. For example, assuming a 10 bit resolution signal quantization, and for a five hidden states

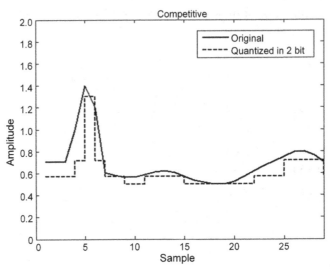

Figure 4.3 Use of competitive quantization to minimize distortion in the amplitude of the digitized signal.

Markov model, the number of parameters to be considered is equal to $1024 \times 5 + 5^2 = 5145$. This high parameter space dimensionality poses severe problems to the estimation algorithms. The volume of the search space increases dramatically with the problem dimensionality leading to problems of data sparsity. That is, the amount of data required for the parameter estimation procedure increases, at an exponential rate, with the problem dimentionality. This fact is a very well known problem in machine learning, and several other data processing sciences, referred to as the "curse of dimensionality".

Due to the computational problems involved when dealing with large dimension problems, a different technique is used when modelling systems with continuous output discurses. This method resorts to the replacement of the emissions probability distribution function by a continuous density function.

To further clarify this concept, let's assume a generic discrete hidden Markov model with m hidden and n observable states, as the one presented in Figure 4.4. Following the concepts introduced during the previous chapter, the coefficients a_{ij} represent the transition probability between hidden states i and j and b_{ij} the probability of ending at state r_j provided that the present active hidden state is s_i. Notice that b_{ij} can be viewed as the

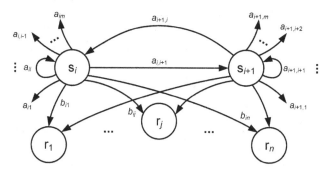

Figure 4.4 Excerpt of a discrete hidden Markov model with n observable states represented from the view point of two hidden states.

outcome of a probability distribution function $b_i(\lambda_k)$ assuming that the active observable state at time k, denoted by λ_k, is r_j. That is, $b_{ij} = b_i(\lambda_k)$ if $\lambda_k = r_j$.

As referred in the previous chapter, this probability distribution function must obey the stochasticity condition. That is, when at hidden state s_i, one of the n observable states must surely become active. This condition can be expressed mathematically as:

$$\sum_{\lambda_k} b_i(\lambda_k) = 1 \tag{4.1}$$

Now, suppose that the number of observable states is increased infinitely. Thus, the above summing expression takes the following integral form,

$$\int_{-\infty}^{+\infty} b_i(\lambda_k)\, d\lambda_k = 1 \tag{4.2}$$

In this context, $b_i(\lambda_k)$ switches from a probability distribution function to a probability density function. The concept of probability density function was introduced in Section 2.3 of Chapter 2.

Besides the stochasticity constraint of $b_i(\lambda_k)$, it is also fundamental to know how the total probability is distributed along the values of the independent variable λ_k. In other words, what is the probability density law underlying the generation of the observations. The shape of the probability density function can be very simple, as in the case of continuous uniform distribution, or can be arbitrarily complex. Assuming that $b_i(\lambda_k)$ has a smooth profile, it is almost always possible to decompose it by means of a weighted sum of simpler basis density probability functions. For example, assuming that $b_i(\lambda_k)$ can be expressed as a linear combination of M basis probability density functions $\mathfrak{D}(\lambda_k, \Theta_i)$ parameterized by the vector Θ_i, then,

$$b_i(\lambda_k) = \sum_{i=1}^{M} w_i \cdot \mathfrak{D}(\lambda_k, \Theta_i) \tag{4.3}$$

for w_i, as the weighting factor associated to the probability density function $\mathfrak{D}(\lambda_k, \Theta_i)$, and meeting the following constraints:

$$\sum_{i=1}^{M} w_i = 1$$
$$w_i \geq 0 \tag{4.4}$$

Note that there are multiple choices for the probability density function $\mathfrak{D}(\lambda_k, \Theta_i)$ used above as basis. Nevertheless, generally it is assumed to be the Gaussian probability distribution function or, in case of counting processes, the Poisson distribution. If the former function is considered, then (4.3) is referred as a Gaussian mixture. The following section will be devoted to further explore this concept.

4.2 PROBABILITY DENSITY FUNCTIONS AND GAUSSIAN MIXTURES

It is fundamental to understand the statistical concept of continuous probability distribution before addressing the continuous distribution hidden Markov models[1]. It is also very important to analyze the condition in which the statistical probability distributions, related to the observations, do not follow standard probability density functions such as the Gaussian or Poisson functions. The case of multimodal probability density functions is one typical example. In those cases, the function can be viewed as being composed by a linear combination of more standard and regular basis functions. Although several different functions can be used as a basis, in this book only Gaussian functions will be considered. In this section, first the case of univariate Gaussian functions will be presented followed by the multivariate situation. For both cases, the effect on the overall function

---

[1]Or "continuous hidden Markov models" for short.

due to changes in the mean and variance of each of the basis Gaussian functions, will be analysed. Finally, it will be discussed how to find the appropriate weights w_i of (4.3) in order to obtain an arbitrary density functions from the linear combination of Gaussian functions. This weight estimation procedure will be first addressed in the context of the univariate problem and then, it will be extrapolated to handle the multivariate case.

4.2.1 GAUSSIAN FUNCTIONS IN SYSTEM MODELING

The Gaussian function has already be presented at (2.41) during Chapter 2. The value of this function depends on the distance of a generic point x to the center of the function, μ, and on a factor that determines the function spread, σ.

This function is very common and can be found in a myriad of different knowledge areas such as physics, statistics, mathematics, machine learning and signal processing, just to name a few. Under the statistics frame of reference, the Gaussian function appears as the mathematical representation of the Normal probability density function: a law that a large number of physical processes seems to follow. For example, imagine you drop, by accident, a glass into the floor of your kitchen and it breaks. The probability of finding glass shards near the drop zone is higher than in more distant regions.

Within the field of signal processing, machine learning and function approximation, the Gaussian can be found as an activation function in radial basis function (RBF) neural networks, as a kernel function in support vector machines (SVM) or even as a membership function in fuzzy models. However, the Gaussian function in the context of hidden Markov models, plays a completely different role when comparing to the three previous examples.

To make this difference more evident to the reader, the following paragraphs provide a brief description of the action performed by the Gaussian functions in RBF neural networks and Takagi-Sugeno fuzzy models.

In the RBF neural network field, there are several distinct types of functions that can be used as the neurons activation function. For example the inverse multi-quadratic function and *splines* are two possible choices. However, the most frequently used function is the symmetric Gaussian. The structure of a RBF neural network involves three layers: an input layer that contains the set of input neurons, the hidden layer which contains the processing neurons, and finally the output layer with the output neurons. Figure 4.5 illustrates this type of model considering, for simplicity, only a single output neuron.

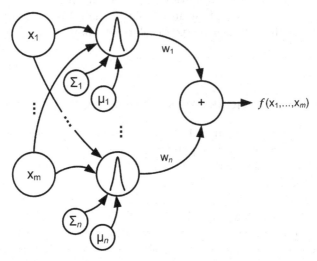

Figure 4.5 A radial-based neural network (RBF-NN) with only one output neuron.

In this type of network, each neuron of the hidden layer has a multivariate Gaussian radial basis function $\mathcal{G}(\mathbf{x}, \boldsymbol{\mu}, \boldsymbol{\Sigma})$ parameterized by $\boldsymbol{\mu}$ and $\boldsymbol{\Sigma}$, and mathematically described by:

$$\mathcal{G}(\mathbf{x}, \boldsymbol{\mu}_i, \boldsymbol{\Sigma}_i) = \exp\left(-(\mathbf{x} - \boldsymbol{\mu}_i)^T \boldsymbol{\Sigma}_i^{-1}(\mathbf{x} - \boldsymbol{\mu}_i)\right) \tag{4.5}$$

where $\boldsymbol{\mu} = [\mu_1 \cdots \mu_d]^T$, for d as the problem dimension, is the means vector and $\boldsymbol{\Sigma}$ is a square, symmetric and positive definite matrix, denoted by the covariance matrix and expressed as:

$$\boldsymbol{\Sigma} = \begin{bmatrix} \sigma^2(x_1) & \cdots & \sigma(x_1, x_d) \\ \vdots & \ddots & \vdots \\ \sigma(x_d, x_1) & \cdots & \sigma^2(x_d) \end{bmatrix} \tag{4.6}$$

where $\sigma(x_i, x_j)$ is the covariance between the component x_i and x_j of the input vector $\mathbf{x}$ for i and j taking values from 1 to d.

Geometrically, $\boldsymbol{\mu}$ represents the center location of the Gaussian function, and $\boldsymbol{\Sigma}$ can be interpreted as the base area that the function occupies in space.

The output neurons perform a weighted average along the set of signals provided by each of the hidden layer neurons. This operation is carried out by multiplying each hidden neuron output signal by a coefficient w_i. That is, the signal at the output neuron is:

$$f(\mathbf{x}) = \sum_{i=1}^{n} w_i \cdot \mathcal{G}(\mathbf{x}, \boldsymbol{\mu}_i, \boldsymbol{\Sigma}_i) \tag{4.7}$$

where n refers to the number of neurons in the hidden layer. The outputs amplitude of the hidden units will be large when the input is near the centroid described by the Gaussian functions. This value will decrease, symmetrically, as the Euclidean distance between the input vector and the neuron centroid increase. Remark that the model output consists on a weighted sum of Gaussian functions. This type of function aggregation is commonly referred as a Gaussian mixture. The Gaussian function parameters, $\boldsymbol{\mu}_i$ and $\boldsymbol{\Sigma}_i$, and each of the weighing coefficients w_i are commonly obtained using a supervising training procedure. During this process, the RBF neural network parameters are iteratively changed according to the training data presented. After the parameters estimation procedure, the artificial neural network model can be used to handle data whose statistical behavior is similar to the one used during the training.

On the other hand, when dealing with Fuzzy based models, the Gaussian function emerges in the shape of membership functions. Their purpose is to execute the partition of the discourse universe into groups. In order to highlight the conceptual differences between a Fuzzy model and the RBF neuronal network, a diagram of the Takagi-Sugeno Fuzzy model is provided in Figure 4.6.

Consider that the input space is divided into n partitions. Then, for an arbitrary input vector $\mathbf{x}$, its membership degree to each one of the n partitions is determined. Let $\phi_i(\mathbf{x})$ be the degree of membership of the input vector $\mathbf{x}$ regarding the space partition i for $i = 1, \cdots, n$. Assuming a Gaussian membership function, then $\phi_i(\mathbf{x}) = \mathcal{G}(\mathbf{x}, \boldsymbol{\mu}_i, \boldsymbol{\Sigma}_i)$. Moreover, let's define $\phi_T(\mathbf{x})$ as the membership average of $\mathbf{x}$:

$$\phi_T(\mathbf{x}) = \sum_{i=1}^{n} \phi_i(\mathbf{x}) \tag{4.8}$$

Then, the model output $f(\mathbf{x})$ will be computed as,

$$f(\mathbf{x}) = \frac{1}{\phi_T(\mathbf{x})} \sum_{i=1}^{n} \phi_i(\mathbf{x})(\theta_i \mathbf{x} + b_i) \tag{4.9}$$

As in the RBF neural network, the Fuzzy model training algorithm will try to adjust the model parameters in order to achieve the best match possible according to the presented training examples. Once again, a Gaussian mixture can be identified within the function presented in (4.9).

Regarding the hidden Markov models, the Gaussian mixture function that will be included in the model, does not have the same purpose as in the previous two cases. Now, instead of being using as mapping from $\mathbf{x}$ into $f(\mathbf{x})$, it will be used to define the profile of a probability density function. In order to further elaborate this statement, consider once again a hidden Markov model with m hidden states and n observable states. Consider that, at the time instant k, the model is at hidden state s_i. It has been seen in the previous chapter that the use of this discrete Markov model involves the following sequence of operations:

1. A number between 1 and m is randomly extracted using a probability distribution defined by the ith row of the state transition matrix $\mathbf{A}$. This operation returns the next active hidden state at instant $k + 1$;
2. The active observable state is obtained using a similar strategy to the one described in the previous item. That is, a random number between 1 and n is extracted with a probability density given by the row i of matrix $\mathbf{B}$;
3. Finally, a mapping between the observable state and the current data target vector is performed.

However, and this is a fundamental question to keep present, in the continuous hidden Markov models, there is no $\mathbf{B}$ matrix. What exists is a function that characterizes the probability density associated to a given hidden state. Even if other approaches are possible, the most usual way to mathematically describe the probability density function is through a linear combination of Gaussian functions (similar to that of a RBF neural network). The number of such functions that are incorporated in the mixture, varies and depends on the current active hidden state.

The first step in running a continuous hidden Markov models is similar to the first step on its discrete counterpart. However, now the observation vector is obtained from a random number generator with a continuous probability distribution function. The shape of this function is defined by means of a linear combination of Gaussian functions. The following section will be devoted to explore how this mixture of Gaussian functions can lead to more complex probability density functions.

166

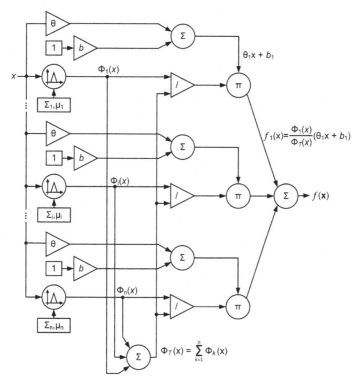

Figure 4.6 Diagram of a fuzzy model with Gaussian membership functions (Takagi-Sugeno model).

4.2.2 GAUSSIAN FUNCTION AND GAUSSIAN MIXTURE

The Gaussian probability density function, associated to a random variable λ, denoted by $\mathcal{G}(\lambda, \mu, \sigma^2)$, has the following algebraic formulation:

$$\mathcal{G}(\lambda, \mu, \sigma^2) = \frac{1}{\sqrt{2\pi\sigma^2}} \exp\left(-\frac{(\lambda - \mu)^2}{2\sigma^2}\right) \qquad (4.10)$$

where μ refers to the distribution mean and σ^2 to its variance. The multiplicative factor $\frac{1}{\sqrt{2\pi\sigma^2}}$ is added in order to force the Gaussian function to obey the stochasticity constraint. That is,

$$\int_{-\infty}^{+\infty} \mathcal{G}(\lambda, \mu, \sigma^2) \, d\lambda = 1 \qquad (4.11)$$

167

In practice, the outcome of many physical experiments, follows a Gaussian probability density function. However, there are situations where this probability distribution function does not accurately fit the observed data. In this type of problems, the probability density function can frequently be described as a linear combination of basis density functions. Let this basis be the Gaussian function. In this case, the global probability density function $p(\lambda|\Theta)$ obtained by selective mixing M distinct Gaussian functions is denoted by:

$$p(\lambda|\Theta) = \sum_{i=1}^{M} w_i \cdot \mathcal{G}(\lambda, \mu_i, \sigma_i^2) \qquad (4.12)$$

where Θ is the parameters set composed by the weights, means and variances associated to each of the M Gaussian functions. That is, $\Theta = \{\Psi_1, \cdots, \Psi_M\}$ for $\Psi_i = \{w_i, \mu_i, \sigma_i^2\}$ and $i = 1, \cdots, M$. Furthermore, the weights w_i must also comply to the set of conditions expressed at (4.4).

In the above expression, the weights w_i can be viewed as the probability of the actual density function to be described by the Gaussian with mean μ_i and variance σ_i^2. Moreover, since $p(\lambda|\Theta)$ is a probability density function, it must also obey to the stochasticity constraint. That is,

$$\int_{-\infty}^{+\infty} p(\lambda|\Theta) \, d\lambda = 1$$
$$p(\lambda|\Theta) \geq 0 \qquad (4.13)$$

Now, if the random variable λ is multidimensional with dimension d and expressed in vector form as $\boldsymbol{\lambda} = [x_1 \cdots x_d]^T$, the Gaussian function presented in (4.10) will take the following alternative multivariate shape:

$$\mathcal{G}(\boldsymbol{\lambda}, \boldsymbol{\mu}, \boldsymbol{\Sigma}) = \frac{1}{\sqrt{\det(2\pi\boldsymbol{\Sigma})}} \exp\left(-\frac{1}{2}(\boldsymbol{\lambda} - \boldsymbol{\mu})^T \boldsymbol{\Sigma}^{-1}(\boldsymbol{\lambda} - \boldsymbol{\mu})\right) \qquad (4.14)$$

which has already been introduced in (4.5) without the scale factor. The spatial location and shape of this function depend on both the vector $\boldsymbol{\mu}$ and the covariance matrix. Figures 4.7, 4.8 and 4.9 present the profile of multivariate Gaussian functions for the special case of 2D observations assuming different values for $\boldsymbol{\mu}$ and $\boldsymbol{\Sigma}$.

The situation presented in Figure 4.7 refers to (4.14) assuming that the covariance is a diagonal matrix. Changing the covariance matrix main diagonal values, the Gaussian suffers a deformation along its main axis.

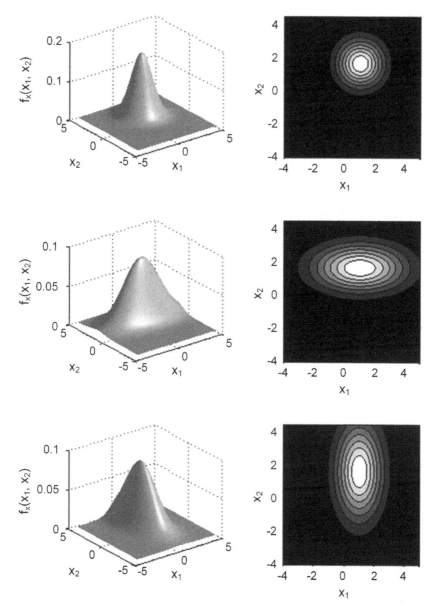

Figure 4.7 Matrix of pure diagonal covariance. (a) Variance of x_1 equal to the variance of x_2. (b) Variance of x_1 four times greater than the variance of x_2. (c) Inverse situation to that presented in (a).

The covariance diagonal element that has the larger value will lead to a greater deformation along the axis associated with this element.

In the sequence of plots presented at Figure 4.8 it is assumed that the covariance matrix is a positive definite matrix. In this framework, both $\sigma^2_{x_1 x_2}$ and $\sigma^2_{x_2 x_1}$ are nonzero and all eigenvalues of $\boldsymbol{\Sigma}$ are positive.

A common feature in all the plots of Figure 4.9 is the fact that the values of the main diagonal of $\boldsymbol{\Sigma}$ were kept constant while the remaining values were changed. Following the images from top to bottom, the effect of increasing the cross-covariance values between x_1 and x_2 is illustrated. It can be seen that the density function is now oriented along a diagonal direction. Moreover, if the values outside the diagonal remain symmetrical but negative, the function "rotates" in the opposite direction.

In the same context as (4.12), but now for the multivariate case, the probability density function obtained by a mixture of M Gaussian functions is expressed as:

$$p(\boldsymbol{\lambda}|\boldsymbol{\Theta}) = \sum_{i=1}^{M} w_i \cdot \mathcal{G}(\boldsymbol{\lambda}, \boldsymbol{\mu}_i, \boldsymbol{\Sigma}_i) \qquad (4.15)$$

where $\boldsymbol{\Theta} = \{\boldsymbol{\Psi}_1, \cdots, \boldsymbol{\Psi}_M\}$ with $\boldsymbol{\Psi}_i = \{w_i, \boldsymbol{\mu}_i, \boldsymbol{\Sigma}_i\}$ for $i = 1, \cdots, M$ and constrained to,

$$\int_{\mathbb{R}^d} p(\boldsymbol{\lambda}|\boldsymbol{\Theta})d\boldsymbol{\lambda} = 1$$
$$p(\boldsymbol{\lambda}|\boldsymbol{\Theta}) \geq 0 \qquad (4.16)$$

where, once again, the weight factors w_i, for $i = 1, \cdots, M$, must obey the set of conditions already presented by (4.4).

Let's assume a generic element λ extracted from a random process whose probability density function consists on the sum of M Gaussian.

This element can be understood as being generated by one of the M basis function where the choice of the Gaussian, that gave rise to it, was made by a random number generator (integers between 1 and M) whose probability distribution is given by $\mathbf{w} = \{w_1, \cdots, w_M\}$. To illustrate this statement, Figure 4.10 shows the profile of a probability density function consisting of a mixture of three univariate Gaussian functions whose centers and standard deviations are $(0, 1)$, $(1, 0.5)$ and $(2, 0.1)$, respectively. The weight of each Gaussian, in the final density function, is $\mathbf{w} = \{0.7, 0.2, 0.1\}$. The same figure presents, superimposed over the mixed probability density function, the normalised histogram composed by a sequence of 5000 values extracted from randomly selecting the outcome of

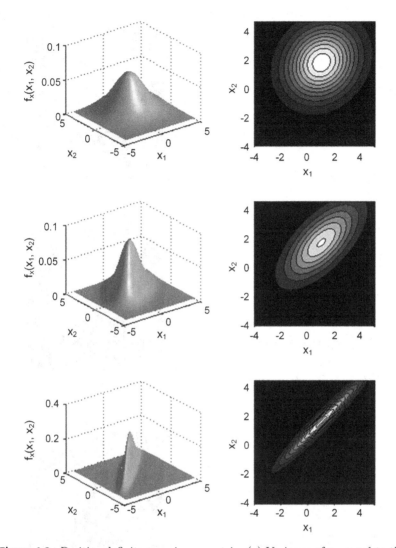

Figure 4.8 Positive-definite covariance matrix. (a) Variance of x_1 equal to the variance of x_2. (b) Variance of x_1 four times greater than the variance of x_2. (c) Inverse situation to that presented in (a).

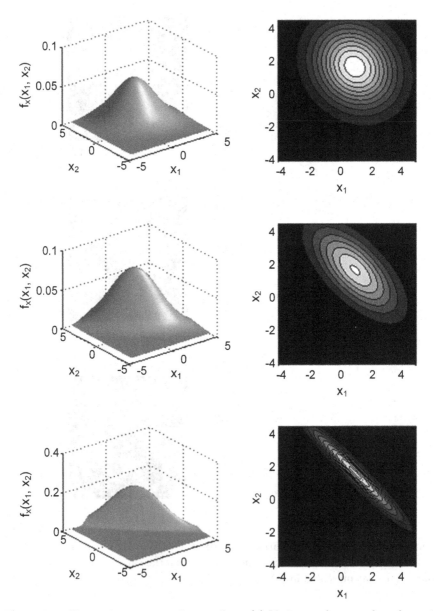

Figure 4.9 Change in cross-covariance values. (a) Variance of x_1 equal to the variance of x_2. (b) Variance of x_1 four times greater than the variance of x_2. (c) Inverse situation to that presented in (a).

one of three distinct random processes according to the distribution function **w**. Those three random processes have Gaussian probability distribution functions, whose parameters are equal to those of the above referred mixture. The MATLAB® code presented in Listing 4.1 was used to generate the plot from Figure 4.1. Note that this function makes a call to the function "random_func()" presented in Listing 3.20 and "gaussian()" presented at Listing 4.2.

Listing 4.1 MATLAB® script for gaussian mixture example.

```
1  % Generate de Gaussian mixture
2  x=-5:.1:4;
3  y1=gaussian(x,0,1);          % Gaussian #1
4  y2=gaussian(x,1,0.5);        % Gaussian #2
5  y3=gaussian(x,2,0.1);        % Gaussian #3
6  w=[0.7 0.2 0.1];             % Relative weights
7
8  yt=w(1)*y1+w(2)*y2+w(3)*y3;  % Gaussian mixture
9
10 % Generate 5000 samples from 3 distinct Gaussian
11 % according to a distribution function w
12 mix=random_func(w,5000);
13 h=zeros(1,5000);
14 for i=1:5000,
15     if mix(i)==1,
16         h(i)=0 + 1.*randn(1,1);
17     elseif mix(i)==2,
18         h(i)=1 + 0.5.*randn(1,1);
19     else
20         h(i)=2 + 0.1.*randn(1,1);
21     end
22 end
23
24 % Present the data
25 plot(x,yt)
26 hold on
27 histogram(h,100,'Normalization','pdf')
```

Listing 4.2 Gaussian function called by the MATLAB® script described in Listing 4.1.

```
1  function N=gaussian(x,mu,sigma)
2  % Gaussian function
3  N=(1/sqrt(2*pi*sigma.^2)).*exp(-0.5*((x-mu)./sigma).^2);
```

A clear coincidence between the histogram and the probability density function is observed. Thus, it is confirmed that an element generated by a mixed probability density function can be seen as the extraction, using

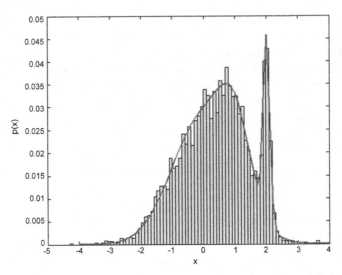

Figure 4.10 A sequence of 5000 values generated by a multimodal distribution. Comparison of the probability density function with the histogram of the generated sequence.

the relative weights as a distribution function, of one of the Gaussian in the mixture.

In practice, the coefficients of the Gaussian mixture are unknown. Only a data set, gathered from a random process with an arbitrary probability density function, is available. The problem is then to estimate the values of the unknown coefficients using the information provided by the data set. It is worth to notice that $p(\lambda|\Theta)$ is parametric in Θ. Hence, the probability density estimation problem tries to search for the set of parameters Θ in order to maximize the likelihood of a given set of N observations, $\Lambda_N = \{\lambda_1, \cdots, \lambda_N\}$, to be generated by a stochastic process with probability density function $p(\lambda|\Theta)$.

Assuming statistical independence between observations, the likelihood $\mathcal{L}(\Theta|\Lambda_N)$ function is computed as:

$$\mathcal{L}(\Theta|\Lambda_N) = \prod_{j=1}^{N} p(\lambda_j|\Theta) \tag{4.17}$$

or alternatively, after applying the logarithm transformation over (4.15), by:

$$\mathcal{L}^e(\Theta|\Lambda_N) = \log\left(\mathcal{L}(\Theta|\Lambda_N)\right)$$

$$= \sum_{j=1}^{N} \log\left(\sum_{i=1}^{M} w_i \cdot \mathcal{G}(\lambda_j, \mu_i, \Sigma_i)\right) \quad (4.18)$$

The parameters Θ can then be chosen, from within the parameters space universe, as the ones that maximize the function $\mathcal{L}^e(\Theta|\Lambda_N)$. For this reason, and taking into consideration the stochastic imposition on the mixture weight coefficients, the parameters Θ can be obtained by solving the following constraint optimization problem:

$$\max_{\Theta} \quad \mathcal{L}^e(\Theta|\Lambda_N)$$

$$s.t. \quad \sum_{i=1}^{M} w_i = 1 \quad (4.19)$$

$$w_i \geq 0$$

Let's start by forming the Lagrangian associated to the above problem which leads to:

$$L(\Theta, \gamma) =$$

$$\sum_{j=1}^{N} \log\left(\sum_{i=1}^{M} w_i \cdot \mathcal{G}(\lambda_j, \mu_i, \Sigma_i)\right) + \gamma\left(\sum_{i=1}^{M} w_i - 1\right) \quad (4.20)$$

where $\gamma \in \mathbb{R}$ is the Lagrange multiplier.

Finding the Lagrangian critical point along the w_l direction, for $l = 1, \cdots, M$, requires the computation of the Lagrangian partial derivative with respect to w_l and set it to zero. That is,

$$\frac{\partial L(\Theta, \gamma)}{\partial w_l} = \sum_{j=1}^{N} \frac{\partial}{\partial w_l} \log\left(\sum_{i=1}^{M} w_i \cdot \mathcal{G}(\lambda_j, \mu_i, \Sigma_i)\right) + \gamma$$

$$= \sum_{j=1}^{N} \frac{\mathcal{G}(\lambda_j, \mu_l, \Sigma_l)}{\sum_{i=1}^{M} w_i \cdot \mathcal{G}(\lambda_j, \mu_i, \Sigma_i)} + \gamma = 0 \quad (4.21)$$

Multiplying and dividing the above expression by w_l yields:

$$\frac{\partial L(\mathbf{\Theta}, \gamma)}{\partial w_l} = \sum_{j=1}^{N} \frac{w_l \cdot \mathcal{G}(\boldsymbol{\lambda}_j, \boldsymbol{\mu}_l, \boldsymbol{\Sigma}_l)}{w_l \cdot \sum_{i=1}^{M} w_i \cdot \mathcal{G}(\boldsymbol{\lambda}_j, \boldsymbol{\mu}_i, \boldsymbol{\Sigma}_i)} + \gamma \qquad (4.22)$$
$$= 0$$

Let's define the posterior probability $p(\mathcal{G}_l | \boldsymbol{\lambda}_j)$ as:

$$p(\mathcal{G}_l | \boldsymbol{\lambda}_j) = \frac{w_l \cdot \mathcal{G}(\boldsymbol{\lambda}_j, \boldsymbol{\mu}_l, \boldsymbol{\Sigma}_l)}{\sum_{i=1}^{M} w_i \cdot \mathcal{G}(\boldsymbol{\lambda}_j, \boldsymbol{\mu}_i, \boldsymbol{\Sigma}_i)} \qquad (4.23)$$

that is, the probability of having a Gaussian density distribution generator process with parameters $\{\boldsymbol{\mu}_l, \boldsymbol{\Sigma}_l\}$ given the observation $\boldsymbol{\lambda}_j$. With this new definition, (4.22) becomes,

$$\frac{\partial L(\mathbf{\Theta}, \gamma)}{\partial w_l} = \sum_{j=1}^{N} \frac{1}{w_l} p(\mathcal{G}_l | \boldsymbol{\lambda}_j) + \gamma = 0 \qquad (4.24)$$

The Lagrange multiplier γ can then be found by taking the above equation and computing the cumulative sum along all the values for l leading to:

$$\sum_{l=1}^{M} \sum_{j=1}^{N} p(\mathcal{G}_l | \boldsymbol{\lambda}_j) = -\gamma \sum_{l=1}^{M} w_l \qquad (4.25)$$

Since both $\sum_{l=1}^{M} w_l$ and $\sum_{l=1}^{M} p(\mathcal{G}_l | \boldsymbol{\lambda}_j)$ are equal to 1 then the above expression resumes to $\gamma = -N$. Using this result into (4.24), and solving for w_l, the following result is obtained:

$$w_l = \frac{1}{N} \sum_{j=1}^{N} p(\mathcal{G}_l | \boldsymbol{\lambda}_j) \qquad (4.26)$$

This last expression can be interpreted as follows: the probability of generating the sequence of observations Λ_N, by a random process with a Gaussian probability distribution function $\mathcal{G}(\boldsymbol{\lambda}, \boldsymbol{\mu}_l, \boldsymbol{\Sigma}_l)$, is equal to the average value obtained from the probability taken along all the observations. Even if (4.26) is intuitive, this expression does not lead to a closed form solution for w_l since, in order to be able to compute $p(\mathcal{G} | \boldsymbol{\lambda}_i)$, it is necessary to know w_l.

Let us now calculate the location of the Lagrangian critical points along $\boldsymbol{\mu}_l$. In order to do this, the probability density function $\mathcal{G}(\lambda_i, \boldsymbol{\mu}_j, \boldsymbol{\Sigma}_j)$ in (4.20) is replaced by expression (4.14) leading to:

$$
L(\mathbf{w}, \mathbf{u}, \mathbf{s}, \gamma) =
$$

$$
\sum_{i=1}^{N} \log \left(\sum_{j=1}^{M} \frac{w_j}{\sqrt{\det(2\pi\boldsymbol{\Sigma}_j)}} \exp\left(-\frac{1}{2} \left(\lambda_i - \boldsymbol{\mu}_j\right)^T \boldsymbol{\Sigma}_j^{-1} \left(\lambda_i - \boldsymbol{\mu}_j\right) \right) \right)
$$

$$
+ \gamma \left(\sum_{i=1}^{M} w_i - 1 \right)
$$

$$
\tag{4.27}
$$

The partial derivative of $L(\mathbf{w}, \mathbf{u}, \mathbf{s}, \gamma)$ in order to $\boldsymbol{\mu}_l$ is:

$$
\frac{\partial}{\partial \boldsymbol{\mu}_l} L(\mathbf{w}, \mathbf{u}, \mathbf{s}, \gamma) =
$$

$$
\sum_{i=1}^{N} \frac{\partial}{\partial \boldsymbol{\mu}_l} \log \left(\sum_{j=1}^{M} \frac{w_j \exp\left(-\frac{1}{2} \left(\lambda_i - \boldsymbol{\mu}_j\right)^T \boldsymbol{\Sigma}_j^{-1} \left(\lambda_i - \boldsymbol{\mu}_j\right) \right)}{\sqrt{\det(2\pi\boldsymbol{\Sigma}_j)}} \right)
\tag{4.28}
$$

which, according to the logarithm derivative rule, leads to:

$$
\frac{\partial}{\partial \boldsymbol{\mu}_l} L(\mathbf{w}, \mathbf{u}, \mathbf{s}, \gamma) =
$$

$$
\sum_{i=1}^{N} \frac{\frac{\partial}{\partial \boldsymbol{\mu}_l} \sum_{j=1}^{M} w_j \exp\left(-\frac{1}{2} \left(\lambda_i - \boldsymbol{\mu}_j\right)^T \boldsymbol{\Sigma}_j^{-1} \left(\lambda_i - \boldsymbol{\mu}_j\right) \right)}{\sqrt{\det(2\pi\boldsymbol{\Sigma}_j)} \sum_{j=1}^{M} w_j \mathcal{G}\left(\lambda_i, \boldsymbol{\mu}_j, \boldsymbol{\Sigma}_j\right)}
\tag{4.29}
$$

Since,

$$
\frac{\partial}{\partial \boldsymbol{\mu}_l} \sum_{j=1}^{M} w_j \exp\left(-\frac{1}{2} \left(\lambda_i - \boldsymbol{\mu}_j\right)^T \boldsymbol{\Sigma}_j^{-1} \left(\lambda_i - \boldsymbol{\mu}_j\right) \right) =
$$

$$
w_l \frac{\partial}{\partial \boldsymbol{\mu}_l} \exp\left(-\frac{1}{2} \left(\lambda_i - \boldsymbol{\mu}_l\right)^T \boldsymbol{\Sigma}_l^{-1} \left(\lambda_i - \boldsymbol{\mu}_l\right) \right) =
\tag{4.30}
$$

$$
w_l \boldsymbol{\Sigma}_l^{-1} \left(\lambda_i - \boldsymbol{\mu}_l\right) \exp\left(-\frac{1}{2} \left(\lambda_i - \boldsymbol{\mu}_l\right)^T \boldsymbol{\Sigma}_l^{-1} \left(\lambda_i - \boldsymbol{\mu}_l\right) \right)
$$

then,

$$\frac{\partial}{\partial \boldsymbol{\mu}_l} L(\mathbf{w}, \mathbf{u}, \mathbf{s}, \gamma) =$$

$$\sum_{i=1}^{N} \frac{\exp\left(-\frac{1}{2}\left(\lambda_i - \boldsymbol{\mu}_l\right)^T \boldsymbol{\Sigma}_l^{-1}\left(\lambda_i - \boldsymbol{\mu}_l\right)\right)}{\sqrt{\det\left(2\pi\boldsymbol{\Sigma}_l\right)}} \frac{w_l \boldsymbol{\Sigma}_l^{-1}\left(\lambda_i - \boldsymbol{\mu}_l\right)}{\sum_{j=1}^{M} w_j \mathcal{G}\left(\lambda_i, \boldsymbol{\mu}_j, \boldsymbol{\Sigma}_j\right)} = \quad (4.31)$$

$$\sum_{i=1}^{N} \boldsymbol{\Sigma}_l^{-1}\left(\lambda_i - \boldsymbol{\mu}_l\right) \frac{w_l \mathcal{G}\left(\lambda_i, \boldsymbol{\mu}_l, \boldsymbol{\Sigma}_l\right)}{\sum_{j=1}^{M} w_j \mathcal{G}\left(\lambda_i, \boldsymbol{\mu}_j, \boldsymbol{\Sigma}_j\right)}$$

which results in:

$$\frac{\partial}{\partial \boldsymbol{\mu}_l} L(\mathbf{w}, \mathbf{u}, \mathbf{s}, \gamma) = \sum_{i=1}^{N} \boldsymbol{\Sigma}_l^{-1}\left(\lambda_i - \boldsymbol{\mu}_l\right) p\left(\mathcal{G}_l | \lambda_i\right) \quad (4.32)$$

Equating the previous expression to zero results in:

$$\sum_{i=1}^{N} \boldsymbol{\Sigma}_l^{-1} \lambda_i \cdot p\left(\mathcal{G}_l | \lambda_i\right) - \sum_{i=1}^{N} \boldsymbol{\Sigma}_l^{-1} \boldsymbol{\mu}_l \cdot p\left(\mathcal{G}_l | \lambda_i\right) = 0$$

$$\boldsymbol{\Sigma}_l^{-1} \boldsymbol{\mu}_l \sum_{i=1}^{N} p\left(\mathcal{G}_l | \lambda_i\right) = \boldsymbol{\Sigma}_l^{-1} \sum_{i=1}^{N} \lambda_i \cdot p\left(\mathcal{G}_l | \lambda_i\right) \quad (4.33)$$

Finally, solving in order to $\boldsymbol{\mu}_l$ leads to the following estimation expression:

$$\boldsymbol{\mu}_l = \frac{\sum_{i=1}^{N} \lambda_i \cdot p\left(\mathcal{G}_l | \lambda_i\right)}{\sum_{i=1}^{N} p\left(\mathcal{G}_l | \lambda_i\right)} \quad (4.34)$$

To obtain the Lagrangian critical point along $\boldsymbol{\Sigma}_l$, the following equation must be solved:

$$\frac{\partial}{\partial \boldsymbol{\Sigma}_l} L(\mathbf{w}, \mathbf{u}, \mathbf{s}, \gamma) = 0 \quad (4.35)$$

that is,

$$\frac{\partial}{\partial \mathbf{\Sigma}_l} L(\mathbf{w}, \mathbf{u}, \mathbf{s}, \gamma) =$$

$$\frac{\partial}{\partial \mathbf{\Sigma}_l} \sum_{i=1}^{N} \log \left(\sum_{j=1}^{M} w_j \frac{\exp\left(-\frac{1}{2}\left(\lambda_i - \boldsymbol{\mu}_j\right)^T \mathbf{\Sigma}_j^{-1}\left(\lambda_i - \boldsymbol{\mu}_j\right)\right)}{\sqrt{\det\left(2\pi\mathbf{\Sigma}_j\right)}} \right)$$

$$+ \frac{\partial}{\partial \mathbf{\Sigma}_l} \gamma \left(\sum_{i=1}^{M} w_i - 1 \right) \tag{4.36}$$

$$\sum_{i=1}^{N} \frac{\partial}{\partial \mathbf{\Sigma}_l} \log \left(\sum_{j=1}^{M} w_j \frac{\exp\left(-\frac{1}{2}\left(\lambda_i - \boldsymbol{\mu}_j\right)^T \mathbf{\Sigma}_j^{-1}\left(\lambda_i - \boldsymbol{\mu}_j\right)\right)}{\sqrt{\det\left(2\pi\mathbf{\Sigma}_j\right)}} \right)$$

The differentiation of the logarithm leads to:

$$\frac{\partial}{\partial \mathbf{\Sigma}_l} \log \left(\sum_{j=1}^{M} w_j \frac{\exp\left(-\frac{1}{2}\left(\lambda_i - \boldsymbol{\mu}_j\right)^T \mathbf{\Sigma}_j^{-1}\left(\lambda_i - \boldsymbol{\mu}_j\right)\right)}{\sqrt{\det\left(2\pi\mathbf{\Sigma}_j\right)}} \right) =$$

$$\frac{\frac{\partial}{\partial \mathbf{\Sigma}_l} \sum_{j=1}^{M} w_j \frac{\exp\left(-\frac{1}{2}\left(\lambda_i - \boldsymbol{\mu}_j\right)^T \mathbf{\Sigma}_j^{-1}\left(\lambda_i - \boldsymbol{\mu}_j\right)\right)}{\sqrt{\det\left(2\pi\mathbf{\Sigma}_j\right)}}}{\sum_{j=1}^{M} w_j \frac{\exp\left(-\frac{1}{2}\left(\lambda_i - \boldsymbol{\mu}_j\right)^T \mathbf{\Sigma}_j^{-1}\left(\lambda_i - \boldsymbol{\mu}_j\right)\right)}{\sqrt{\det\left(2\pi\mathbf{\Sigma}_j\right)}}} \tag{4.37}$$

where the numerator derivative is:

$$\frac{\partial}{\partial \mathbf{\Sigma}_l} \sum_{j=1}^{M} w_j \frac{\exp\left(-\frac{1}{2}\left(\lambda_i - \boldsymbol{\mu}_j\right)^T \mathbf{\Sigma}_j^{-1}\left(\lambda_i - \boldsymbol{\mu}_j\right)\right)}{\sqrt{\det\left(2\pi\mathbf{\Sigma}_j\right)}} =$$

$$\sum_{j=1}^{M} w_j \frac{\partial}{\partial \mathbf{\Sigma}_l} \left(\frac{\exp\left(-\frac{1}{2}\left(\lambda_i - \boldsymbol{\mu}_j\right)^T \mathbf{\Sigma}_j^{-1}\left(\lambda_i - \boldsymbol{\mu}_j\right)\right)}{\sqrt{\det\left(2\pi\mathbf{\Sigma}_j\right)}} \right) = \tag{4.38}$$

$$w_l \frac{\partial}{\partial \mathbf{\Sigma}_l} \left(\frac{\exp\left(-\frac{1}{2}\left(\lambda_i - \boldsymbol{\mu}_l\right)^T \mathbf{\Sigma}_l^{-1}\left(\lambda_i - \boldsymbol{\mu}_l\right)\right)}{\sqrt{\det\left(2\pi\mathbf{\Sigma}_l\right)}} \right)$$

and,

$$\frac{\partial}{\partial \mathbf{\Sigma}_l} \left(\frac{\exp\left(-\frac{1}{2}\left(\lambda_i - \boldsymbol{\mu}_l\right)^T \mathbf{\Sigma}_l^{-1} \left(\lambda_i - \boldsymbol{\mu}_l\right)\right)}{\sqrt{\det\left(2\pi\mathbf{\Sigma}_l\right)}} \right) =$$

$$\frac{\frac{\partial}{\partial \mathbf{\Sigma}_l}\left(\exp\left(-\frac{1}{2}\left(\lambda_i - \boldsymbol{\mu}_l\right)^T \mathbf{\Sigma}_l^{-1} \left(\lambda_i - \boldsymbol{\mu}_l\right)\right)\right) \sqrt{\det\left(2\pi\mathbf{\Sigma}_l\right)}}{\left(\sqrt{\det\left(2\pi\mathbf{\Sigma}_l\right)}\right)^2} - \qquad (4.39)$$

$$\frac{\exp\left(-\frac{1}{2}\left(\lambda_i - \boldsymbol{\mu}_l\right)^T \mathbf{\Sigma}_l^{-1} \left(\lambda_i - \boldsymbol{\mu}_l\right)\right) \frac{\partial}{\partial \mathbf{\Sigma}_l}\sqrt{\det\left(2\pi\mathbf{\Sigma}_l\right)}}{\left(\sqrt{\det\left(2\pi\mathbf{\Sigma}_l\right)}\right)^2}$$

The derivative of the exponential function in the previous expression can be computed as[2]:

$$\frac{\partial}{\partial \mathbf{\Sigma}_l}\left(\exp\left(-\frac{1}{2}\left(\lambda_i - \boldsymbol{\mu}_l\right)^T \mathbf{\Sigma}_l^{-1} \left(\lambda_i - \boldsymbol{\mu}_l\right)\right)\right) =$$

$$-\frac{1}{2}\frac{\partial \left(\lambda_i - \boldsymbol{\mu}_l\right)^T \mathbf{\Sigma}_l^{-1} \left(\lambda_i - \boldsymbol{\mu}_l\right)}{\partial \mathbf{\Sigma}_l} \exp\left(-\frac{1}{2}\left(\lambda_i - \boldsymbol{\mu}_l\right)^T \mathbf{\Sigma}_l^{-1} \left(\lambda_i - \boldsymbol{\mu}_l\right)\right) =$$

$$\frac{1}{2}\mathbf{\Sigma}_l^{-1}\left(\lambda_i - \boldsymbol{\mu}_l\right)\left(\lambda_i - \boldsymbol{\mu}_l\right)^T \mathbf{\Sigma}_l^{-1} \exp\left(-\frac{1}{2}\left(\lambda_i - \boldsymbol{\mu}_l\right)^T \mathbf{\Sigma}_l^{-1} \left(\lambda_i - \boldsymbol{\mu}_l\right)\right)$$

$$(4.40)$$

In addition,

$$\frac{\partial}{\partial \mathbf{\Sigma}_l}\sqrt{\det\left(2\pi\mathbf{\Sigma}_l\right)} = \frac{1}{2}\frac{1}{\sqrt{\det\left(2\pi\mathbf{\Sigma}_l\right)}}\frac{\partial}{\partial \mathbf{\Sigma}_l}\det\left(2\pi\mathbf{\Sigma}_l\right) \qquad (4.41)$$

Since $\det\left(2\pi\mathbf{\Sigma}_l\right) = 2\pi^d \det\left(\mathbf{\Sigma}_l\right)$ where d refers to matrix $\mathbf{\Sigma}_l$ dimension, then $\frac{\partial}{\partial \mathbf{\Sigma}_l}\left[2\pi^d \det\left(\mathbf{\Sigma}_l\right)\right] = 2\pi^d \det\left(\mathbf{\Sigma}_l\right)\mathbf{\Sigma}_l^{-1} = \det\left(2\pi\mathbf{\Sigma}_l\right)\mathbf{\Sigma}_l^{-1}$. Thus,

$$\frac{\partial}{\partial \mathbf{\Sigma}_l}\sqrt{\det\left(2\pi\mathbf{\Sigma}_l\right)} = \frac{\det\left(2\pi\mathbf{\Sigma}_l\right)\mathbf{\Sigma}_l^{-1}}{2\sqrt{\det\left(2\pi\mathbf{\Sigma}_l\right)}} \qquad (4.42)$$

---

[2]Note that $\frac{\partial \mathbf{a}^T \mathbf{A}^{-1}\mathbf{b}}{\partial \mathbf{A}} = -\left(\mathbf{A}^{-1}\right)^T \mathbf{a}\mathbf{b}^T \left(\mathbf{A}^{-1}\right)^T$ for $\mathbf{A}$ an invertible matrix and two vectors of appropriate dimensions $\mathbf{a}$ and $\mathbf{b}$. If, additionally, $\mathbf{A}$ is a symmetric matrix, then $\frac{\partial \mathbf{a}^T \mathbf{A}^{-1}\mathbf{b}}{\partial \mathbf{A}} = -\mathbf{A}^{-1}\mathbf{a}\mathbf{b}^T \mathbf{A}^{-1}$.

Replacing (4.40) and (4.42) into (4.39) and factorizing the exponential function results in:

$$
\frac{\partial}{\partial \boldsymbol{\Sigma}_l} \left(\frac{\exp\left(-\frac{1}{2}\left(\lambda_i - \boldsymbol{\mu}_l\right)^T \boldsymbol{\Sigma}_l^{-1}\left(\lambda_i - \boldsymbol{\mu}_l\right)\right)}{\sqrt{\det\left(2\pi\boldsymbol{\Sigma}_l\right)}} \right) =
$$

$$
\frac{\left(\frac{1}{2}\boldsymbol{\Sigma}_l^{-1}\left(\lambda_i - \boldsymbol{\mu}_l\right)\left(\lambda_i - \boldsymbol{\mu}_l\right)^T \boldsymbol{\Sigma}_l^{-1}\sqrt{\det\left(2\pi\boldsymbol{\Sigma}_l\right)} - \frac{\det\left(2\pi\boldsymbol{\Sigma}_l\right)\boldsymbol{\Sigma}_l^{-1}}{2\sqrt{\det\left(2\pi\boldsymbol{\Sigma}_l\right)}}\right)}{\left(\sqrt{\det\left(2\pi\boldsymbol{\Sigma}_l\right)}\right)^2} \qquad (4.43)
$$

$$
\cdot \exp\left(-\frac{1}{2}\left(\lambda_i - \boldsymbol{\mu}_l\right)^T \boldsymbol{\Sigma}_l^{-1}\left(\lambda_i - \boldsymbol{\mu}_l\right)\right)
$$

That is,

$$
\frac{\partial}{\partial \boldsymbol{\Sigma}_l} \left(\frac{\exp\left(-\frac{1}{2}\left(\lambda_i - \boldsymbol{\mu}_l\right)^T \boldsymbol{\Sigma}_l^{-1}\left(\lambda_i - \boldsymbol{\mu}_l\right)\right)}{\sqrt{\det\left(2\pi\boldsymbol{\Sigma}_l\right)}} \right) =
$$

$$
\frac{\left(\frac{1}{2}\boldsymbol{\Sigma}_l^{-1}\left(\lambda_i - \boldsymbol{\mu}_l\right)\left(\lambda_i - \boldsymbol{\mu}_l\right)^T \boldsymbol{\Sigma}_l^{-1}\sqrt{\det\left(2\pi\boldsymbol{\Sigma}_l\right)} - \frac{\det\left(2\pi\boldsymbol{\Sigma}_l\right)\boldsymbol{\Sigma}_l^{-1}}{2\sqrt{\det\left(2\pi\boldsymbol{\Sigma}_l\right)}}\right)\mathcal{G}\left(\lambda_i, \boldsymbol{\mu}_l, \boldsymbol{\Sigma}_l\right)}{\sqrt{\det\left(2\pi\boldsymbol{\Sigma}_l\right)}} =
$$

$$
\frac{1}{2}\left(\boldsymbol{\Sigma}_l^{-1}\left(\lambda_i - \boldsymbol{\mu}_l\right)\left(\lambda_i - \boldsymbol{\mu}_l\right)^T - 1\right)\boldsymbol{\Sigma}_l^{-1}\mathcal{G}\left(\lambda_i, \boldsymbol{\mu}_l, \boldsymbol{\Sigma}_l\right)
$$

$$
\qquad (4.44)
$$

Therefore,

$$
\frac{\partial}{\partial \boldsymbol{\Sigma}_l} \log \left(\sum_{j=1}^{M} w_j \frac{\exp\left(\left(\lambda_i - \boldsymbol{\mu}_j\right)^T \boldsymbol{\Sigma}_j^{-1}\left(\lambda_i - \boldsymbol{\mu}_j\right)\right)}{\sqrt{\det\left(2\pi\boldsymbol{\Sigma}_j\right)}} \right) =
$$

$$
\frac{1}{2}\frac{\left(\boldsymbol{\Sigma}_l^{-1}\left(\lambda_i - \boldsymbol{\mu}_l\right)\left(\lambda_i - \boldsymbol{\mu}_l\right)^T \boldsymbol{\Sigma}_l^{-1} - \boldsymbol{\Sigma}_l^{-1}\right) w_l \mathcal{G}\left(\lambda_i, \boldsymbol{\mu}_l, \boldsymbol{\Sigma}_l\right)}{\sum_{j=1}^{M} w_j \mathcal{G}\left(\lambda_i, \boldsymbol{\mu}_j, \boldsymbol{\Sigma}_j\right)} = \qquad (4.45)
$$

$$
\frac{1}{2}\left(\boldsymbol{\Sigma}_l^{-1}\left(\lambda_i - \boldsymbol{\mu}_l\right)\left(\lambda_i - \boldsymbol{\mu}_l\right)^T \boldsymbol{\Sigma}_l^{-1} - \boldsymbol{\Sigma}_l^{-1}\right) \cdot p\left(\mathcal{G}_l | \lambda_i\right)
$$

Finally, the Lagrangian derivative in order to $\boldsymbol{\Sigma}_l$, is:

$$
\frac{\partial}{\partial \boldsymbol{\Sigma}_l} L(\mathbf{w}, \mathbf{u}, \mathbf{s}, \gamma) =
$$

$$
\sum_{i=1}^{N} \frac{1}{2}\left(\boldsymbol{\Sigma}_l^{-1}\left(\lambda_i - \boldsymbol{\mu}_l\right)\left(\lambda_i - \boldsymbol{\mu}_l\right)^T \boldsymbol{\Sigma}_l^{-1} - \boldsymbol{\Sigma}_l^{-1}\right) \cdot p\left(\mathcal{G}_l | \lambda_i\right) \qquad (4.46)
$$

Equating to zero and solving for $\boldsymbol{\Sigma}_l$ results in:

$$\boldsymbol{\Sigma}_l = \frac{\sum_{i=1}^{N} p\left(\mathcal{G}_l|\lambda_i\right)\left(\lambda_i - \boldsymbol{\mu}_l\right)\left(\lambda_i - \boldsymbol{\mu}_l\right)^T}{\sum_{i=1}^{N} p\left(\mathcal{G}_l|\lambda_i\right)} \qquad (4.47)$$

In conclusion, the procedure for estimating the parameters $\boldsymbol{\Theta}$ of Gaussian mixture equations, is governed by (4.26), (4.34) and (4.47). It is important to note that, these equations do not lead to a closed-form solution since the calculation of each parameter depends on $p(\mathcal{G}_l|\lambda_i)$ which, in turn, requires the knowledge of $\boldsymbol{\Theta}$. Thus, the procedure is iterative starting from an initial estimate of the solution $\boldsymbol{\Theta}_0$. Using these initial estimates of the parameters, the value of $p(\mathcal{G}_l|\lambda_i)$ is computed. This probability will then be used to obtain new estimates for $\boldsymbol{\Theta}$ using the three equations already mentioned.

This iterative procedure, coded for MATLAB®, can be found in Listing 4.3. Notice that this main function makes calls to Listings 4.4 and 4.5.

Listing 4.3 Definition of PDF with Gaussian mixture: training algorithm.

```
1  function [sol_W,sol_Mu,sol_Sigma]=TrainMdG(O,M,MaxIter)
2  % O - Vector with observations
3  % M - Number of Gaussians
4  [D,N]=size(O); % Number of observations + problem dimension
5  P=zeros(M,N);
6  p=zeros(1,N);
7  % Initialize data containers
8  LogLik=zeros(MaxIter,1);
9  sol_W=zeros(MaxIter,M);
10 sol_Mu=zeros(MaxIter,M,D);
11 sol_Sigma=zeros(MaxIter,M,D,D);
12 Sigma=zeros(M,D,D);
13 Mu=zeros(M,D);
14 % .................................set initial values
15 for j=1:M
16     Sigma(j,:,:)=3*rand(D,D).*eye(D,D);
17     Mu(:,j)=randn(D,1);
18 end
19 W=rand(1,M);W=W/sum(W);
20 % ...................................iterative procedure
21 for iter=1:MaxIter
22     LogLik(iter)=sum(log(Gfit(O,W,Mu,Sigma)));
23     disp('..................................................');
24     disp(['Iteration' num2str(iter-1) ': logLik = ' ...
25         num2str(LogLik(iter))]);
26     sol_W(iter,:)=W;
27     sol_Mu(iter,:,:)=Mu;
28     sol_Sigma(iter,:,:,:)=Sigma;
29     % determine P(Gj/xi)
30     for i=1:N,
31         p(i)=0;
```

```
32              for j=1:M
33                  p(i)=p(i)+W(j)*G(O(:,i),Mu(:,j),Sigma(j,:,:));
34              end
35              for j=1:M,
36                  P(j,i)=W(j)*G(O(:,i),Mu(:,j),Sigma(j,:,:))/p(i);
37              end
38          end
39          % compute new estimates for Mu, Sigma and W
40          %.............................................W
41          for l=1:M
42              soma=0;
43              for i=1:N,
44                  soma=soma+P(l,i);
45              end
46              W(l)=(1/N)*soma;
47          end
48          %.............................................Mu
49          for l=1:M
50              num=0;
51              den=0;
52              for i=1:N,
53                  num=num+O(:,i)*P(l,i);
54                  den=den+P(l,i);
55              end
56
57              Mu(:,l)=num./den;
58          end
59          %.............................................Sigma
60          for l=1:M
61              num=0;
62              den=0;
63              for i=1:N,
64                  num=num+P(l,i)*((O(:,i)-Mu(:,l))*(O(:,i)-Mu(:,l))');
65                  den=den+P(l,i);
66              end
67              Sigma(l,:,:)=num./den;
68          end
69      end
```

Listing 4.4 Function used to compute the value of an uni/multivariate Gaussian function.

```
1 function y=G(x,Mu,Sigma)
2 % Compute uni/multivariate Gaussian
3 L=size(Sigma);
4 if length(L)>2,
5     Sigma=reshape(Sigma,L(2),L(3));
6     y=(1/(sqrt(det(2*pi*Sigma))))*exp(-0.5*(((x-Mu)'/(Sigma))*(x-Mu)));
7 else
8     y=(1/(sqrt(det(2*pi*Sigma))))*exp(-0.5*(((x-Mu)'/(Sigma))*(x-Mu)));
9 end
```

Listing 4.5 Function used to compute the probability density function.

```
1  function pdf=Gfit(x,W,Mu,Sigma)
2  %Compute probability density function
3  L=size(Sigma);
4  [D,M]=size(Mu);
5  if D>1,
6      for k=1:length(x);
7          pdf(k)=0;
8          for j=1:M
9              S1=reshape(Sigma(j,:,:),L(2),L(3));
10             M1=Mu(:,j);
11             pdf(k)=pdf(k)+W(j)/(sqrt(det(2*pi*S1)))*exp(-0.5*(((x(:,k)-
                  M1)'/S1)*(x(:,k)-M1)));
12         end
13     end
14 else
15     for k=1:length(x);
16         pdf(k)=0;
17         for j=1:M
18             S1=Sigma(j);
19             M1=Mu(j);
20             pdf(k)=pdf(k)+W(j)/(sqrt(2*pi*S1))*exp(-0.5*(x(:,k)-M1).^2/
                  S1);
21         end
22     end
23 end
```

This MATLAB® function has been used to generate the plot of Figure 4.11 which represents the parameters estimation problem for a mixture of two Gaussian functions: one with mean 1 and variance 0.1 and the other with mean 2 and variance equal to 0.2. The former will have a relative weight of 0.4 and the latter of 0.6.

Those parameters are assumed unknown and the overall Gaussian mixture function is used to generate a data set composed of 2000 samples. The first part of the MATLAB® script presented in Listing 4.6 was used to generate the data set.

Then, using this data and resorting to the function presented in Listing 4.3, the means, variances and relative weights of the two Gaussian were obtained. The initial parameters estimation was set randomly and the procedure was executed along one thousand iterations.

The final plots were carried out by running the MATLAB® function presented in Listing 4.7 which requires the function in Listing 4.8 in order to properly operate.

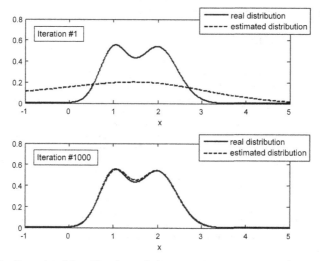

Figure 4.11 Iterative identification of the parameters associated to a mixture of two Gaussian from a sequence of 2000 observations. Above the initial estimation and below the result after 1000 iterations.

Listing 4.6 MATLAB® script used to test the Gaussian mixture parameter estimation problem.

```
1  clear all;
2  close all;
3  clc;
4  %--------------------------------Generate Training Data
5  dx=.01;
6  xx=-1:dx:5;
7  m=[1 2];      % Gaussian Means
8  s=[.1 .2];    % Gaussian Variances
9  d=[0.4 0.6];  % Weighting factors
10 % Generate a training vector with this PDF
11 N=2000;       % Number of samples
12 x=zeros(1,N);% Traing samples vector
13 for k=1:N,
14     inx=rand;
15     if inx<d(1);
16         x(k)=m(1)+sqrt(s(1))*randn;
17     else
18         x(k)=m(2)+sqrt(s(2))*randn;
19     end
20 end
21 % Use of EM to estimate the Gaussian mixture
22 Maxiter=1000;   % Number of iterations
23 M=2;            % Number of Gaussian functions in the mixture
24 %--------------------------------Training Algorithm
25 [sol_W,sol_Mu,sol_Sigma]=TrainMdG(x,M,Maxiter);
26 %---------------------------------------------------
27 plotEM(d,m,s,sol_W,sol_Mu,sol_Sigma);
```

Listing 4.7 Code used to generate the Figure 4.11.

```
1  function plotEM(W,Mu,Sigma,sol_W,sol_Mu,sol_Sigma)
2  [~,pdf1]=drawGauss(W,Mu,Sigma);
3  figure
4  subplot(2,1,1);
5  [xx,pdf2]=drawGauss(sol_W(1,:),sol_Mu(1,:),sol_Sigma(1,:));
6  plot(xx,pdf1,xx,pdf2); axis([-1 5 0 0.8]);xlabel('x');
7  legend('real distribution','estimated distribution');
8  text(-0.8,0.7,'Iteration #1');
9  subplot(2,1,2);
10 [xx,pdf2]=drawGauss(sol_W(end,:),sol_Mu(end,:),sol_Sigma(end,:));
11 plot(xx,pdf1,xx,pdf2);axis([-1 5 0 0.8]);xlabel('x');
12 legend('real distribution','estimated distribution');
13 text(-0.8,0.7,'Iteration #1000');
```

Listing 4.8 Generate data do draw the Gaussian probability function.

```
1  function [xx,pdf]=drawGauss(W,Mu,Sigma)
2  dx=.01;
3  xx=-1:dx:5;
4  pdf=(W(1)/(sqrt(2*pi*Sigma(1))))*exp(-0.5*((xx-Mu(1)).^2)/Sigma(1))+(W
      (2)/(sqrt(2*pi*Sigma(2))))*exp(-0.5*((xx-Mu(2)).^2)/Sigma(2));
```

The previous example illustrates the behavior of the training algorithm assuming an one dimensional problem. However, it is possible to apply the same algorithm for a generic d-dimensional problem. The top left image of Figure 4.12 presents a multimodal probability density function that is intended to be approximated by a mixture of two multivariate Gaussian functions. The MATLAB® code used to generate it is provided in Listing 4.9.

Five hundred observations are generated by means of this probability density distribution. The function presented in Listing 4.10, for N=500, u1=[1;2], v1=[.1 0;0 .2], u2=[2;1] and v2=[.3 0;0 .2], shows how this can be generated.

Listing 4.9 MATLAB® script for plotting the top left 2D density function illustrated in Figure 4.12.

```
1  dx = .1;                        % Space resolution
2  x  = -1:dx:4;                   % Space limits
3  N = [length(x) length(x)];      % Number of points
4  [x1,x2]=meshgrid(x);            % Create mesh
5  xx=[x1(:) x2(:)]';              % 2D data for plot
6  Nt = length(xx);                % Number of points
7  u1=[1;2];                       % Mean for Gauss. 1
8  v1=[.1 0;0 .2];                 % Cov. for Gauss. 1
9  u2=[2;1];                       % Mean for Gauss. 2
```

```
10  v2=[.3 0;0 .2];              % Cov. for Gauss. 2
11  w =[0.4 0.6];                % Mixing weights
12  pdf = zeros(1,Nt);           % Mixed pdf
13  %-------------------------------------------------
14  for j=1:Nt;
15      pdf(j)=w(1)/(sqrt(det(2*pi*v1)))*...
16      exp(-0.5*(((xx(:,j)-u1)'/v1)*(xx(:,j)-u1)))+...
17      w(2)/(sqrt(det(2*pi*v2)))*...
18      exp(-0.5*(((xx(:,j)-u2)'/v2)*(xx(:,j)-u2)));
19  end
20  %-------------------------------------------------
21  x1=reshape(xx(1,:),N);
22  x2=reshape(xx(2,:),N);
23  pdf=reshape(pdf,N);
24  %-------------------------------------------------
25  surfl(x1,x2,pdf)             % Plot PDF
26  shading interp              %
27  colormap gray               %
28  axis tight                  %
29  xlabel('x_1');ylabel('x_2');  %
```

Listing 4.10 MATLAB® function to generate the 2D data set.

```
1  function x=generate_2D_MoG_Data(u1,v1,u2,v2,w,N)
2  x=zeros(2,N);                % Training examples matrix
3  for k=1:N,
4      if rand<w(1);
5          x(:,k)=u1+sqrt(v1)*randn(2,1);
6      else
7          x(:,k)=u2+sqrt(v2)*randn(2,1);
8      end
9  end
```

The next step is to iteratively obtain the parameters for the Gaussian mixture. The MATLAB® code presented in Listing 4.11 does exactly this. Additionally, this script requires the function described in Listing 4.12 in order to graphically present the results which, in turn, requires the functions presented in Listing 4.13.

Listing 4.11 MATLAB® script to obtain the parameter estimation of a 2D Gaussian mixture function.

```
1  u1=[1;2];                    % means
2  u2=[2;1];
3  v1=[.1 0;0 .2];              % variances (positive definite)
4  v2=[.3 0;0 .2];
5  w=[0.4 0.6];                 % weights
6  % Generate with this pdf a training vector:
7  N=500;                       % Number of samples
8  M=2;
9  MaxIter=100;
```

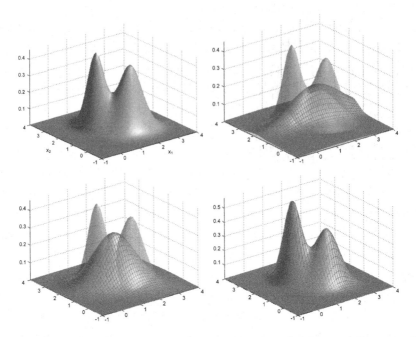

Figure 4.12 Gaussian mixture parameters estimation. At the top left corner the original density function. At the top right the initial estimation. At the bottom left the result after one iteration. At the bottom right the result after 150 iterations.

```
10  x=generate_2D_MoG_Data(u1,v1,u2,v2,w,N);
11  %--------------------------------------------Training Algorithm
12  [sol_W,sol_Mu,sol_Sigma]=TrainMdG(x,M,MaxIter);
13  %-------------------------------------------------------------
14  v(1,:,:)=v1;v(2,:,:)=v2;
15  plotEM2(w,u,v,sol_W,sol_Mu,sol_Sigma);
```

Listing 4.12 MATLAB® function to graphically present the results of the parameters estimation procedure.

```
1  function plotEM2(W,Mu,Sigma,sol_W,sol_Mu,sol_Sigma)
2  [~,pdf]=drawGauss2(W,Mu,Sigma);
3  figure;
4  [xx,pdf2]=drawGauss2(sol_W(1,:),sol_Mu(1,:,:),sol_Sigma(1,:,:,:));
5  N1=sqrt(length(xx));
6  x1=reshape(xx(1,:),N1,N1);
7  x2=reshape(xx(2,:),N1,N1);
8  pdf=reshape(pdf,N1,N1);
9  pdf2=reshape(pdf2,N1,N1);
10 h=surfl(x1,x2,pdf);
```

```
11  set(h,'FaceAlpha',0.4);
12  shading interp
13  hold on;
14  h=surfl(x1,x2,pdf2);
15  shading interp
16  colormap gray;
17  set(h,'EdgeColor',[0.5 0.5 0.5]);
18  axis tight
19  %------------------------------------------------------Sub Function
20  [xx,pdf2]=
21  drawGauss2(sol_W(end,:),sol_Mu(end,:,:),sol_Sigma(end,:,:,:));
22  figure;
23  N1=sqrt(length(xx));
24  x1=reshape(xx(1,:),N1,N1);
25  x2=reshape(xx(2,:),N1,N1);
26  pdf=reshape(pdf,N1,N1);
27  pdf2=reshape(pdf2,N1,N1);
28  h=surfl(x1,x2,pdf);
29  set(h,'FaceAlpha',0.4);
30  shading interp
31  hold on;
32  h=surfl(x1,x2,pdf2);
33  shading interp
34  colormap gray;
35  set(h,'EdgeColor',[0.5 0.5 0.5]);
36  axis tight
```

Listing 4.13 MATLAB® function called by the one presented in Listing 4.11.

```
1   function [xx,pdf]=drawGauss2(W,Mu,Sigma)
2   L=size(Sigma);
3   if length(L)>3
4       v1=reshape(Sigma(1,1,:,:),L(3),L(4));
5       v2=reshape(Sigma(1,2,:,:),L(3),L(4));
6   else
7       v1=reshape(Sigma(1,:,:),L(2),L(3));
8       v2=reshape(Sigma(2,:,:),L(2),L(3));
9   end
10  L=size(Mu);
11  if length(L)>2
12      Mu=reshape(Mu(1,:,:),L(2),L(3));
13  end
14  dx=.1;
15  [x1,x2]=meshgrid(-1:dx:4);
16  xx=[x1(:) x2(:)]';
17  for k=1:length(xx);
18      pdf(k)=W(1)/(sqrt(det(2*pi*v1)))*exp(-0.5*(((xx(:,k)-Mu(:,1))'/v1)
            *(xx(:,k)-Mu(:,1))))+...
19          W(2)/(sqrt(det(2*pi*v2)))*exp(-0.5*(((xx(:,k)-Mu(:,2))'/v2)*(xx
            (:,k)-Mu(:,2))));
20  end
```

The estimation process begins by performing an initial random estimation for Θ. Special care must be taken in order to ensure that the covariance matrices are positive definite. The initial estimated Gaussian

189

mixture has the shape illustrated in the top right image of Figure 4.12. Then, for each iteration, the *a posteriori* probability $p(\mathcal{G}_i|\boldsymbol{\lambda}_i)$ is computed followed by new estimation of $\boldsymbol{\mu}_i$, $\boldsymbol{\Sigma}_i$ and w_i for $i = 1$ to 2. The bottom left and bottom right plots of Figure 4.12 represent the estimated Gaussian mixture after 1 and 150 iterations, respectively.

Before ending this section, it is important to highlight some facts about this Gaussian mixture estimation procedure. First of all, the algorithm efficiency strongly depends on the amount and quality of the estimation data. Additionally, the random parameters initialization used above can be replaced by a more convenient clustering procedure. The "k-means" and CURE algorithms are good examples of such techniques [32, 1]. And finally, by not being a closed form solution, and due to the search space non-convexity, this method only guarantees convergence to a local optimum.

4.3 CONTINUOUS HIDDEN MARKOV MODEL DYNAMICS

In continuous distribution hidden Markov models, $b_i(\lambda_k)$ is replaced by a probability density function. This function is usually described by means of a linear Gaussian mixtures. This is due to the fact that mixture of Gaussian functions are able to approximate, with arbitrary precision, any multimodal continuous function [20].

In this section, the temporal dynamics of a continuous hidden Markov model will be addressed. In order to do this, observe from Figure 4.13 a partial structure of a continuous hidden Markov model where the observations probability density functions are mixtures taken from M Gaussian functions. Note the presence of feedback between the output of the probability density function $b_i(\lambda_k)$, and the probability of observing, from a given hidden state the value λ_k. That is, $b_i(\lambda_k)$ reappears in the loop connecting the hidden state s_i to the observable state λ_k. This feature can be interpreted as follows: due to the infinite number of observable states, the transition probability from the hidden state s_i to an arbitrary observable state λ_k, is determined as the value of the probability density function taken for λ_k.

For the generic case, where the observations are multidimensional, then λ_k is equal to $[x_{1k} \cdots x_{dk}]^T$, where the subscript d denotes the observations dimensionality. Now, let's assume a set of N d-dimensional observations represented by $\Lambda_N = \{\boldsymbol{\lambda}_1, \cdots, \boldsymbol{\lambda}_N\}$. Additionally, for each m hidden state, and for every one of the M Gaussian functions, the covariance matrix $\boldsymbol{\Sigma}$, the means vector $\boldsymbol{\mu}$ and the weights w are defined. In

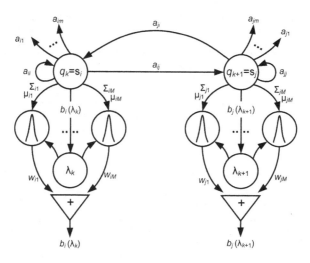

Figure 4.13 Continuous distribution hidden Markov model. Partial structure by considering only the transition from hidden state s_i to s_j.

particular, for the ith hidden state and the jth Gaussian on the mixture, the covariance matrix has the following shape:

$$\boldsymbol{\Sigma}_{ij}^k = \begin{bmatrix} \sigma_{ij}^2(x_{1k}) & \cdots & \sigma_{ij}(x_{1k}, x_{dk}) \\ \vdots & \ddots & \vdots \\ \sigma_{ij}(x_{dk}, x_{1k}) & \cdots & \sigma_{ij}^2(x_{dk}) \end{bmatrix} \tag{4.48}$$

Let's also define $\mathbf{V}_i = \{\boldsymbol{\Sigma}_{i1}, \cdots, \boldsymbol{\Sigma}_{iM}\}$ as the set of all the covariance matrices associated to the hidden state i.

This notation can also be extended to both the Gaussian mixture centers and weights. Regarding the former, the mean vector associated to the ith hidden state, and for the jth observation, is denoted by:

$$\boldsymbol{\mu}_{ij} = [\mu_{ij}(1) \cdots \mu_{ij}(d)]^T \tag{4.49}$$

Furthermore, let $\mathbf{U}_i = \{\boldsymbol{\mu}_{i1}, \cdots, \boldsymbol{\mu}_{iM}\}$ be the set of all mean vectors associated to the hidden state i. Finally, the weights associated to hidden state i are organized in vector format as $\mathbf{w}_i = [w_{i1} \cdots w_{iM}]$. In a more compact format, the set of parameters associated to the hidden layer i is represented by $\boldsymbol{\Theta}_i = \{\mathbf{w}_i, \mathbf{U}_i, \mathbf{V}_i\}$.

At this point, the main objective is to find a method that is able to provide estimations for all those model parameters. In the same way as for the discrete hidden Markov model, this procedure can be carried out by the Baum-Welch method. However, due to the probability density function introduction the forward and backward probabilities are computed by slightly different approaches.

The following section will be dedicated to present the conceptual differences in the computation of both the forward and backward probabilities. Additionally, the Viterbi algorithm will also be covered.

4.3.1 FORWARD, BACKWARD AND VITERBI ALGORITHMS REVISITED

This section begins by recalling the definition of the forward probability $\alpha_k(i)$. This quantity concerns the probability of observing a sequence $\Lambda_k = \{\lambda_1 \cdots \lambda_k\}$ and, simultaneously, the active hidden state at time instant k to be s_i, for $i = 1, \cdots, m$. That is,

$$\alpha_k(i) = P\left(\Lambda_k \wedge q_k = s_i\right) \tag{4.50}$$

As discussed in Section 3.2.1, the forward probability value can be efficiently generated, by a recursive computation procedure, starting from:

$$\alpha_1(i) = b_i(\lambda_1) \cdot c_i, \ 1 \leq i \leq m \tag{4.51}$$

where c_i regards the probability of having s_i as the active hidden state when at $k = 1$.

Then, the forward probability is subsequently updated, as a function of k, by the following equation:

$$\alpha_k(i) = b_i(\lambda_k)\left(\sum_{j=1}^{m} a_{ji} \cdot \alpha_{k-1}(j)\right) \tag{4.52}$$

for $1 \leq k \leq N$ and $1 \leq i \leq m$.

However, for continuous hidden Markov models, the observations are no longer characterized by a discrete probability distribution. Instead, the observations probability is described by the density function $p(\lambda|\Theta)$.

For this reason, and assuming a priori knowledge about $\mathbf{U}$, $\mathbf{V}$ and $\mathbf{w}$ for each one of the hidden states, the recursive equations presented in (4.51) and (4.52) are modified as:

$$\alpha_1(i) = p(\lambda_1|\Theta_i) \cdot c_i \tag{4.53}$$

$$\alpha_k(i) = p(\boldsymbol{\lambda}_k|\boldsymbol{\Theta}_i) \cdot \left(\sum_{j=1}^{m} a_{ji} \cdot \alpha_{k-1}(j) \right) \tag{4.54}$$

for $1 \leq k \leq N$ and $1 \leq i \leq m$.

In short, for continuous hidden Markov models, the calculation of the forward probability first requires the computation, for each hidden state, of the probability density $p(\boldsymbol{\lambda}_k|\boldsymbol{\Theta}_i)$. This computation depends upon the knowledge about the parameters $\mathbf{U}$, $\mathbf{V}$ and $\mathbf{w}$ associated to each hidden state. Then, the ordinary forward algorithm is executed by replacing $b_i(\lambda_k)$ by $p(\boldsymbol{\lambda}_k|\boldsymbol{\Theta}_i)$.

Remark that the direct computation of $\alpha_k(i)$, by the above equations, can lead to numeric instability due to the fact that it requires the successive product of values smaller than one. This fact can be avoided by applying a normalization step between successive iterations. The MATLAB® code presented in Listing 4.14 can be used to compute the forward probabilities taking into consideration the above mentioned normalization step.

Listing 4.14 The continuous observations forward algorithm.

```
1  function [Alfa,LogLik]=forward_continuous_norm(A,B,c)
2  % Forward algorithm with normalization
3  %[Alfa,LogLik]=forward_continuous_norm(A,B,O,c)
4  % m hidden states, n output states and N observations
5  % A - mxm (state transitions matrix)
6  % B - nxm (confusion matrix)
7  % c - 1xm (priors vector)
8
9  [m,N]=size(B);
10 Alfa=zeros(N,m);
11 u=zeros(1,N);
12
13 % Initialization
14 for k=1:m
15     Alfa(1,k)=c(k)*B(k,1);
16 end
17
18 % Scaling coefficient
19 u(1)=1/(sum(Alfa(1,:)));
20 Alfa(1,:)=u(1)*Alfa(1,:);
21
22 % Recursion
23 for l=2:N,
24     for k=1:m,
25         S=0;
26         for i=1:m,
27             S=S+A(i,k)*Alfa(l-1,i);
28         end
29         Alfa(l,k)=B(k,1)*S;
30     end
```

193

```
31      u(1)=1/(sum(Alfa(1,:)));
32      Alfa(1,:)=u(1)*Alfa(1,:);
33 end
34
35 % Compute Log Likelihood
36 LogLik=-sum(log(u));
```

Following the same line of thought, the backward probability, denoted by $\beta_k(i)$, is the probability of getting a fraction of the observable sequence from a given hidden state. This probability is also iteratively computed from:

$$\beta_{k-1}(i) = \sum_{j=1}^{m} b_j(\boldsymbol{\lambda}_k) \cdot a_{ij} \cdot \beta_k(j) \tag{4.55}$$

where, $i = 1, \ldots, m$, $k = 1, \ldots, N$ and $\beta_N(i) = 1$, for $i = 1, \cdots, m$.

Since, for continuous hidden Markov models the probability of getting a particular observation $\boldsymbol{\lambda}_k$ is defined by $p(\boldsymbol{\lambda}|\boldsymbol{\Theta}_i)$, the computation of $\beta_k(i)$ is then carried out by:

$$\beta_{k-1}(i) = \sum_{j=1}^{m} p(\boldsymbol{\lambda}_k|\boldsymbol{\Theta}_j) \cdot a_{ij} \cdot \beta_k(j) \tag{4.56}$$

The implementation of the backward algorithm requires, first of all, the computation of $p(\boldsymbol{\lambda}|\boldsymbol{\Theta})$ for a given observation and for all hidden states. Then, equation (4.56) is iterated up to $k = 2$.

Observe that this algorithm also suffers from the same numeric stability problem as the forward algorithm. Hence, a normalisation step between successive iterations must be performed. Listing 4.15 presents a MATLAB® function that can be used to compute the backward probability.

Listing 4.15 The backward algorithm continuous observations version.

```
1 function Beta=backward_continuous_norm(A,B)
2 % Backward algorithm with normalization
3 %[Beta]=backward_continuous_norm(A,B)
4 % m hidden states, n output states and N observations
5 % A - mxm (state transitions matrix)
6 % B - nxm (confusion matrix)
7
8 [m,N]=size(B);
9
10 %% Initialization
11 Beta=zeros(N,m);
12 for k=1:m
```

```
13        Beta(N,k)=1;
14   end
15   u(N)=1/sum(Beta(N,:));
16
17   %% Recursion
18   for t=N-1:-1:1,
19       for i=1:m,
20            Beta(t,i)=0;
21            for j=1:m,
22                 Beta(t,i)=Beta(t,i)+A(i,j)*B(j,t+1)*Beta(t+1,j);
23            end
24       end
25       u(t)=1/sum(Beta(t,:));
26       Beta(t,:)=u(t)*Beta(t,:);
27   end
```

This section ends with a very brief reference to the Viterbi algorithm. For further insights on this algorithm please refer to [31]. As discussed in Section 3.2.3, the Viterbi algorithm can be used to compute the likeliest hidden states sequence. When applied to continuous hidden Markov models, the Viterbi algorithm requires the computation of the emissions probabilities by using the observations values and the Gaussian parameters Θ (a step also common to both the forward and backward algorithms). The MATLAB® function provided in Listing 4.16 can be used to compute the more probable hidden states sequence $q_1, \cdots, q_N$ using the Viterbi algorithm.

Listing 4.16 The MATLAB® function for Viterbi algorithm applied to continuous hidden Markov models.

```
1  function q=viterbi_continuous(A,B,c)
2  % Viterbi algorithm
3  % q=viterbi_continuous(A,B,c)
4  % m hidden states, n output states and N observations
5  % A - mxm (state transitions matrix)
6  % B - nxm (confusion matrix)
7  % c - 1xm (priors vector)
8
9  [m,N]=size(B);
10
11 %% Initialization
12 delta=zeros(N,m);phi=zeros(N,m);
13 tmp=zeros(m,1);q=zeros(N,1);
14 t=1;
15 for k=1:m
16      delta(t,k)=c(k)*B(k,t);
17      phi(t,k)=0;
18 end
19 delta(t,:)=delta(t,:)/sum(delta(t,:));
20
21 %% Recursion
22 for t=2:N,
```

```
23      for k=1:m,
24          for l=1:m,
25              tmp(l)=delta(t-1,l)*A(l,k)*B(k,t);
26          end
27          [delta(t,k),phi(t,k)]=max(tmp);
28      end
29      delta(t,:)=delta(t,:)/sum(delta(t,:));
30  end
31
32  %% Path finding
33  [~,Inx]=max(delta(N,:));
34  q(N)=Inx;
35  for k=N-1:-1:1,
36      q(k)=phi(k+1,q(k+1));
37  end
```

Just before ending this section, Listing 4.17 provides a *testbench* script that can be used for testing the above three algorithms. In particular, a two hidden states, two Gaussian mixture density function, is used over a data set obtained from a random Gaussian probability function process with zero mean and unity variance. Notice that the MATLAB® function presented in Listing 4.18, will be used to compute the emissions probabilities that will be later used by all the three algorithms.

Listing 4.17 A script testbench for testing the forward backward and Viterbi algorithms.

```
1  clear all;
2  clc;
3
4  m=2;                        % Number of hidden states
5  d=3;                        % Observations dimension
6  Lambda=randn(3,10);         % Observations (taken from a
7                              % random process with zero
8                              % mean and unity variance)
9  W=[0.6 0.4;0.5 0.5];        % Weight vector
10 A=[0.7 0.3;0.2 0.8];        % State transition matrix
11 c=[0.5;0.5];                % Initial distribution
12                             %
13 mu(1,1,:)=[-1;1;1];         % The Gaussian's parameters
14 mu(1,2,:)=[2;-2;0];         % 
15 mu(2,1,:)=[-2;-1;0];        % 
16 mu(2,2,:)=[0;-1;0];         % 
17
18 Sigma(1,1,:,:)=diag([0.2 0.3 0.1]);
19 Sigma(1,2,:,:)=diag([1 0.5 2]);
20 Sigma(2,1,:,:)=diag([0.1 0.1 0.1]);
21 Sigma(2,2,:,:)=diag([1 2 3]);
22
23 % Computation of the observations probabilities
24 B=phi(Lambda,mu,Sigma,W);
25
26 % Forward Algorithm
```

```
27 [Alpha,LogLik] = forward_continuous_norm(A,B,c);
28
29 % Backward Algorithm
30 Beta = backward_continuous_norm(A,B);
31
32 % Viterbi Algorithm
33 q = viterbi_continuous(A,B,c);
```

Listing 4.18 MATLAB® function required to compute $p(\lambda|\Theta)$.

```
 1 function B=phi(Lambda,mu,Sigma,W)
 2 % Compute phi
 3 % B=phi(Lambda,mu,Sigma,W)
 4
 5 [m,M,~,~]=size(Sigma);
 6 [d,N]=size(Lambda);
 7 B=zeros(m,N);
 8
 9 % Iteration along each hidden state
10 for i=1:m
11     % Iteration along each observation
12     for j=1:N
13         B(i,j)=0;
14         % Iteration along each Gaussian
15         for k=1:M
16             S=reshape(Sigma(i,k,:,:),d,d);
17             V=reshape(mu(i,k,:),d,1);
18             B(i,j)=B(i,j) + W(i,k) *(1/sqrt(det(2 * pi * S))) * ...
19                 exp(-0.5 * ((Lambda(:,j) - V)'/S) *(Lambda(:,j) - V));
20         end
21     end
22 end
```

The next section focus on the Baum-Welch training method and on the adaptations that must be carried out to make it available to continuous distributions hidden Markov models.

4.4 CONTINUOUS OBSERVATIONS BAUM-WELCH TRAINING AL-GORITHM

The Baum-Welch method has already been extensively discussed in Section 3.3.2. During the current section, this parameter estimation method will be extended to handle the continuous observations hidden Markov model.

The Baum-Welch method is an iterative procedure that starts from an initial estimation for both the model parameters and structure. Additionally, for continuous observation hidden Markov models, the number of

197

functions in the Gaussian mixture, along with its parameters, must also be initially assumed for each hidden state.

When compared to the original Baum-Welch method, this new version requires the replacement of the observation probabilities matrix by a probability density function. The same strategy was done previously in the forward, backward and Viterbi algorithms.

The probability density function is generically described by a mixture of M Gaussian functions. Hence, for each hidden state i, the probability density function is given by:

$$p(\boldsymbol{\lambda}|\boldsymbol{\Theta}_i) = \sum_{j=1}^{M} w_{ij} \cdot \mathcal{G}_{ij} \qquad (4.57)$$

where, for sake of simplicity, $\mathcal{G}_{ij}$ is just a more compact notation for $\mathcal{G}(\boldsymbol{\lambda}, \boldsymbol{\mu}_{ij}, \boldsymbol{\Sigma}_{ij})$.

Assuming a N observations sequence $\boldsymbol{\Lambda}_N$, the probability of the jth component, of the ith mixture, to have generated the observation $\boldsymbol{\lambda}_k$ is denoted by:

$$\rho_k(i,j) = p\left(q_k = s_i \wedge h_{ik} = \mathcal{G}_{ij}|\boldsymbol{\Lambda}_N\right) \qquad (4.58)$$

where h_{ik} regards the active Gaussian, from the universe of the M possible ones, over the mixture associated to hidden state s_i at instant k.

Taking into consideration the Bayes theorem, the above expression can be rewritten as:

$$\begin{aligned} &p\left(q_k = s_i \wedge h_{ik} = \mathcal{G}_{ij}|\boldsymbol{\Lambda}_N\right) \\ &= p\left(h_{ik} = \mathcal{G}_{ij}|q_k = s_i \wedge \boldsymbol{\Lambda}_N\right) \cdot p(q_k = s_i|\boldsymbol{\Lambda}_N) \end{aligned} \qquad (4.59)$$

where $p(h_{ik} = \mathcal{G}_{ij}|q_k = s_i \wedge \boldsymbol{\Lambda}_N)$ can be understood as the probability of generating $\boldsymbol{\lambda}_k$ by the jth Gaussian component of $p(\boldsymbol{\lambda}|\boldsymbol{\Theta}_i)$. On the other hand, $p(q_k = s_i|\boldsymbol{\Lambda}_N)$ denotes the probability, given the sequence $\boldsymbol{\Lambda}_N$ of having, at instant k, s_i as the active hidden state. Let this probability be represented by $\gamma_k(i)$. Then,

$$\rho_k(i,j) = \gamma_k(i) \cdot p(h_{ik} = \mathcal{G}_{ij}|q_k = s_i \wedge \boldsymbol{\Lambda}_N) \qquad (4.60)$$

Figure 4.14 represents the structure of a continuous observation hidden Markov model given s_i as the present active hidden state. At this time instant, the observation $\boldsymbol{\lambda}_k$ is drawn from the jth Gaussian function $\mathcal{G}_{ij}$ with probability w_{ij}.

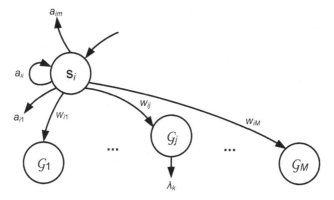

Figure 4.14 Given the active hidden state s_i, illustration of the probability of observing λ_k by the Gaussian function $\mathcal{G}_{ij}$.

After careful examination of this figure, it becomes straightforward to see that the probability $p(h_{ik} = \mathcal{G}_{ij}|q_k = s_i \wedge \Lambda_N)$ is derived from the ratio:

$$p(h_{ik} = \mathcal{G}_{ij}|q_k = s_i \wedge \Lambda_N) = \frac{w_{ij}\mathcal{G}_{ij}^k}{\sum_{l=1}^{M} w_{il}\mathcal{G}_{il}^k} \tag{4.61}$$

where $\mathcal{G}_{ij}^k$ denotes the value of the Gaussian function $\mathcal{G}_{ij}$ taken at $\lambda = \lambda_k$. That is $\mathcal{G}_{ij}^k = \mathcal{G}(\lambda_k, \boldsymbol{\mu}_{ij}, \boldsymbol{\Sigma}_{ij})$.

Replacing (4.61) into (4.60) yields:

$$\rho_k(i,j) = \gamma_k(i) \cdot \frac{w_{ij} \cdot \mathcal{G}_{ij}^k}{\sum_{l=1}^{M} w_{il} \cdot \mathcal{G}_{il}^k} \tag{4.62}$$

Using the above definition for $\rho_k(i,j)$, the Baum-Welch estimation equations for $\hat{w}_{ij}$, $\hat{\boldsymbol{\mu}}_{ij}$ and $\hat{\boldsymbol{\Sigma}}_{ij}$ are described by [23]:

$$\hat{w}_{ij} = \frac{\sum_{k=1}^{N} \rho_k(i,j)}{\sum_{k=1}^{N} \sum_{l=1}^{M} \rho_k(i,l)} \tag{4.63}$$

$$\hat{\boldsymbol{\mu}}_{ij} = \frac{\sum_{k=1}^{N} \rho_k(i,j) \cdot \lambda_k}{\sum_{k=1}^{N} \rho_k(i,j)} \tag{4.64}$$

$$\hat{\boldsymbol{\Sigma}}_{ij} = \frac{\sum_{k=1}^{N} \rho_k(i,j) \left(\lambda_k - \hat{\boldsymbol{\mu}}_{ij} \right) \left(\lambda_k - \hat{\boldsymbol{\mu}}_{ij} \right)^T}{\sum_{k=1}^{N} \rho_k(i,j)} \qquad (4.65)$$

The probability $\rho_k(i,j)$, transversal to all the three above expressions, is computed by the MATLAB® function presented at Listing 4.19.

Listing 4.19 Compute $\rho_k(ij)$.

```
 1  function rho=rho_continuous(O,Gamma,Mu,Sigma,W)
 2  % Compute rho
 3  % rho=rho_continuous(O,Gamma,Mu,Sigma,W)
 4
 5  [N,~]=size(Gamma);
 6  [m,M]=size(W);
 7  G=compute_G(O,Mu,Sigma);
 8  rho=zeros(N,m,M);
 9
10  for k=1:N,
11      for i=1:m;
12          C=0;
13          for l=1:M,
14              C=C+W(i,l)*G(i,l,k);
15          end
16          for j=1:M,
17              rho(k,i,j)=Gamma(k,i)*W(i,j)*G(i,j,k)/(C+(C==0));
18          end
19      end
20  end
```

Regarding (4.65), and in order to avoid the singularity of $\hat{\boldsymbol{\Sigma}}_{ij}$, usually a regularization term ϵ is added to slightly disturb the elements of its main diagonal. That is,

$$\hat{\boldsymbol{\Sigma}}_{ij} = \hat{\boldsymbol{\Sigma}}_{ij} + \epsilon \qquad (4.66)$$

where ϵ can be, for example, $\epsilon \cdot \mathbf{I}$ being $\mathbf{I}$ an identity matrix with same dimension of $\hat{\boldsymbol{\Sigma}}_{ij}$ and ϵ a very small scalar perturbation value.

Due to the similarity between (4.63)-(4.65) and the expressions derived at Section 4.2 for the Gaussian mixtures, it is possible to conclude that they were obtained using a similar procedure. Also, note the resemblance between the probability $\rho_k(i,j)$, for a given hidden state s_i, and the probability $p(\mathcal{G}_j|\boldsymbol{\lambda}_k)$ presented in the same section.

The hidden states probability transitions are obtained using the same approach as for the discrete hidden Markov models. The initial probability estimation for hidden state s_i, denoted by $\hat{c}_i$, and the transition probability between hidden states s_i and s_j, expressed by a_{ij} for $\{i,j\} = 1, \cdots, m$, are computed by:

$$\hat{c}_i = \gamma_1(i) \qquad (4.67)$$

$$\hat{a}_{ij} = \frac{\tau_{ij}}{\tau_i} \qquad (4.68)$$

where τ_{ij} and τ_i are:

$$\tau_{ij} = \sum_{k=1}^{N-1} \xi_k(i,j) \qquad (4.69)$$

$$\tau_i = \sum_{k=1}^{N-1} \gamma_k(i) \qquad (4.70)$$

for $\xi_k(i,j)$ and $\gamma_k(i)$ defined by:

$$\gamma_k(i) = \frac{\alpha_k(i)\,\beta_k(i)}{\sum_{i=1}^{m} \alpha_k(i)\,\beta_k(i)} \qquad (4.71)$$

$$\xi_k(i,j) = \frac{\alpha_k(i)\,a_{ij}b_j(\lambda_{k+1})\,\beta_{k+1}(j)}{\sum_{i=1}^{m} \alpha_k(i)\,\beta_k(i)} \qquad (4.72)$$

where $\alpha_k(i)$ and $\beta_k(i)$ are obtained by the continuous versions of the forward and backward algorithms respectively. Equations (4.69) and (4.70) are coded as MATLAB® functions and presented in Listing 4.20 and Listing 4.21, respectively.

Listing 4.20 Compute τ_{ij}.

```
1  function tau=tau_continuous(Alfa,Beta,A,B)
2  % Compute tau
3  % tau=tau_continuous(Alfa,Beta,A,B)
4
5  [m,N]=size(B);
6  tau=zeros(m,m);
7
8  for k=1:N-1,
9      tmp1=A.*(Alfa(k,:).'*Beta(k+1,:)).*(B(:,k+1)*ones(1,m)).';
10     tmp2=ones(1,m)*tmp1*ones(m,1);
11     tau=tau+tmp1/tmp2;
12 end
```

Listing 4.21 Compute τ_i.

```
1  function taui=taui_continuous(Gamma,B)
2  % Compute taui
3  % taui=taui_continuous(Gamma,B)
4
5  [m,N]=size(B);
6  taui=Gamma(1:N-1,:);
7  taui=(sum(taui,1)).'*ones(1,m);
```

All the above estimation equations are wrapped up in the MAT-LAB® function at Listing 4.22, which also depends on the two functions presented in Listings 4.23 and 4.24. This function is executed with an initial estimation of $\mathbf{A}$, $\boldsymbol{\mu}$, $\boldsymbol{\Sigma}$, $\mathbf{w}$ and $\mathbf{c}$. Additionally, the maximum iterations number must be provided.

Listing 4.22 The Baum-Welch algorithm for continuous observations hidden Markov models.

```
1  function [A,Mu,Sigma,W,c]=baum_welch_cont(O,A,Mu,Sigma,W,c,MaxIter)
2  % Baum-Welch method for continuous HMM
3  % [A,Mu,Sigma,W,c]=baum_welch_cont(O,A,Mu,Sigma,W,c,MaxIter)
4
5  [D,N]=size(O);[m,M]=size(W);
6
7  % Cycle while iterations <= MaxIter
8  for iter=1:MaxIter
9      % Compute B
10     B=phi(O,Mu,Sigma,W); %(ok)
11     % Compute ALFA using the forward algorithm
12     [Alfa,~]=forward_continuous_norm(A,B,c);
13     % Iteration info.
14     disp(['Iteration - ' num2str(iter) '(Lik = ' num2str(LogLik) ')']);
15     % Compute BETA using the backward algorithm
16     Beta=backward_continuous_norm(A,B);
17     % Compute Gama
18     Gamma=compute_gamma(Alfa,Beta);
19     % Compute Tau
20     tau=tau_continuous(Alfa,Beta,A,B);
21     % Compute Taui
22     taui=taui_continuous(Gamma,B);
23     % Estimation of initial probability c
24     c=Gamma(1,:);
25     % Estimation of probability transition matrix A
26     A=tau./taui;
27     % Compute Rho
28     rho=rho_continuous(O,Gamma,Mu,Sigma,W);
29     % Estimation of weights W
30     for i=1:m,
31         for j=1:M,
32             num=0;
33             for k=1:N,
34                 num=num+rho(k,i,j);
35             end
36             den=0;
```

```
37                     for  k=1:N,
38                         for  l=1:M
39                             den=den+rho(k,i,l);
40                         end
41                     end
42                     W(i,j)=num/den;
43             end
44     end
45     % Estimation of mean vector Mu
46     for i=1:m,
47         for j=1:M,
48             num=0;
49             den=0;
50             for k=1:N,
51                 num=num+rho(k,i,j)*O(:,k);
52                 den=den+rho(k,i,j);
53             end
54             Mu(i,j,:)=num/den;
55         end
56     end
57     % Estimation of covariance Sigma
58     for i=1:m,
59         for j=1:M,
60             num=0;
61             den=0;
62             for k=1:N,
63                 M1=reshape(Mu(i,j,:),D,1);
64                 num=num+rho(k,i,j)*(O(:,k)-M1)*(O(:,k)-M1)';
65                 den=den+rho(k,i,j);
66             end
67             Sigma(i,j,:,:)=num/den;
68         end
69     end
70     % Add regularization term to Sigma
71     id=eye(D,D);e=zeros(m,M,D,D);
72     for i=1:m,
73         for j=1:M,
74             e(i,j,:,:)=0.01*id;
75         end
76     end
77     Sigma=Sigma+e;
78 end
```

Listing 4.23 Compute $\gamma_k(i)$.

```
1 function Gamma=compute_gamma(Alfa,Beta)
2 % Compute gamma
3 % Gamma=compute_gamma(Alfa,Beta)
4
5 [~,m]=size(Alfa);
6 P=diag(Alfa*Beta')*ones(1,m);
7 Gamma=(Alfa.*Beta)./P;
```

Listing 4.24 Compute $\mathcal{G}_{ij}$.

```
1  function G=compute_G(O,Mu,Sigma)
2  % Compute G
3  % G=compute_G(O,Mu,Sigma)
4
5  [~,N]=size(O);
6  [m,M,D,~]=size(Sigma);
7  G = zeros(m,M,N);
8
9  for k=1:M
10     for i=1:m
11         S1=reshape(Sigma(i,k,:,:),D,D);
12         M1=reshape(Mu(i,k,:),D,1);
13         for j=1:N,
14             G(i,k,j)=(1/(sqrt(det(2*pi*S1))))*exp(-0.5*(((O(:,j)-M1)'/(
                   S1))*...
15                 (O(:,j)-M1)));
16         end
17     end
18 end
```

In Chapter 6, this parameter estimation method will be used to train three continuous observations hidden Markov models in order to classify cardiotocography signals into three different classes: "normal", "suspect" and "pathological".

4.5 SUMMARY

Discrete hidden Markov models cannot easily handle processes whose discourse alphabet is large. This condition is even more severe for multidimensional observations where each dimension is represented on its own continuous domain.

This limitation can be circumvented by replacing the discrete emission probabilities by a continuous density function. Usually this probability density function is approximated by a linear combination of basis functions. A typical choice for this basis is the Gaussian function. This strategy leads to what is conventionally called by a Gaussian mixture function.

This Chapter started by showing how the Gaussian mixture can be used as a strategy to approximate arbitrary continuous distribution functions. Moreover, it has been presented a way to estimate the Gaussian mixture coefficients by solving a constraint optimization problem. The obtained estimation equations were not in closed form leading to the development of an iterative method. A MATLAB® example was provided where arbitrary distribution functions were fitted into data. First it was considered a 1D probability distribution function and then a 2D.

This way to characterize arbitrary probability density functions was used to describe continuous observations in hidden Markov models. The coefficients of this type of model can be obtained, for example, by the Baum-Welch method. This method relies on the efficient computation of both the forward and backward probabilities. Besides their mathematical background, the Baum-Welch method, the forward and backward algorithms, and the Viterbi algorithm were shown implemented in MATLAB®.

5 Autoregressive Markov Models

"He who controls the past controls the future."

–George Orwell

The hidden Markov model refers to a general modelling paradigm that can be applied to many types of stochastic processes. However, by default, it lacks a time structure. That is, for a particular time instant, the observation depends only on the current active hidden state but is conditionally independent of the observation sequence generated in the past. For this reason, and in cases where intrinsic time dependence among observations must be incorporated, a different hidden Markov model approach is required. This alternative solution is commonly known as autoregressive hidden Markov model since it integrates, into its structure, a Markov chain and a set of autoregressive filters. This last element is responsible for the temporal dependence between successive observations.

The aim of this chapter is to present both the mathematical and computational formulation for autoregressive hidden Markov models. First, an overview of its structure will be presented followed by the definition of the likelihood function for autoregressive models. Then, the parameters estimation of this class of models will be addressed. Moreover, since the autoregressive Markov model is mainly used to describe time-series, the way to perform both short-term and long-term predictions will be presented.

5.1 INTRODUCTION

During the system identification procedure associated with an unknown stochastic process, it is frequently observed that a particular dynamic model does not have the same approximation quality within the process

207

execution range. Specially when the system has multiple and distinct operating regimes.

In these cases, instead of having just one, it would be desirable to have a set of different dynamical models, one for each system regime, and a model switching mechanism. This switching mechanism would be responsible for sensing changes in the system regime and to automatically select the best available model from the set.

With this approach, besides the common difficulties encountered when dealing with the single model system identification, there is now an additional problem. To know how, and when, the process regime changes. That is, the points at which the process must switch between models. This switching mechanism is not directly observable and its action can only be estimated from the system present and past output observations. In particular, this book addresses the case where the regime switching mechanism is described by a Markov chain. Each state in the chain encapsulates a different model which is responsible for describing the process at a given regime. Within each regime, the system behaviour is captured by an autoregressive model structure. The model, obtained by binding both a Markov chain and a set of autoregressive models, will be denoted by autoregressive Markov model or ARMM for short.

In the literature, two main implementation strategies for ARMM can be found: one that requires the time-series segmentation and the other applied to the raw observations. The former was initially proposed by [21, 22] and subsequently extended by [23]. The latter, coined as Markov switching models, was formulated in the early nineties of the twentieth century by [11, 12].

The segmentation approach requires the observations to be sliced into a set of distinct sections. Each one of those sections are assigned to a different Markov chain state. Associated to each state, a different autoregressive (AR) model is considered. The parameters, for each particular AR model, are estimated using only the data segment associated with the respective Markov state. The main drawback of this method lies in the process of data partition. On one hand, the data partition must be done empirically and, on the other hand, for a low number of observations per segment, the information carried out within the signal may be insufficient for proper estimation of the parameters.

Conversely, the strategy described in [11] does not require any manual data segmentation. The AR model parameters associated to each Markov chain state are computed using a weighting method. That is, the observations with higher probability of being generated by a given state associated model will have higher weight when compared to the remaining. This estimation strategy is similar to the one used in parameter estimation by the weighted least squares method.

In this book, only this ARMM strategy will be presented. The following sections are devoted to its mathematical description, structure and handle the details regarding the model's operation and parameters estimation.

5.2 ARMM STRUCTURE

In an ARMM, the observations are no longer conditionally independent between each other such as in discrete or continuous observations hidden Markov models. Instead, the sequence of observations is now intertwined and it is assumed that the correlation between observations can be described by autoregressive type models.

This means that the observation at time instant k, denoted by λ_k, is assumed to be generated by a parametric AR model using the previous p observations. It is fundamental to note that this time structure between observations leads to a severe violation of one of the fundamental hidden Markov properties. In particular, the following conditional probability becomes invalid:

$$P(\lambda_k | \lambda_1 \cdots \lambda_{k-1} \wedge q_1 \cdots q_k) = P(\lambda_k | q_k) \tag{5.1}$$

where $q_1, \cdots, q_k$ denotes the active hidden states sequence along the considered time interval. This condition reveals that, for computing the probability of observing λ_k, it is irrelevant to have the knowledge of both the past observations history and the hidden states sequence up to time k.

In ARMM, this independence condition between observations and past states does not hold. For this reason, there are many situations where the results obtained for hidden Markov models cannot be applied. For example, the forward, backward and Viterbi algorithms can no longer be used within the ARMM reference frame.

Before presenting the basic structure of an autoregressive Markov model, let's assume the following p-order autoregressive process which generates some observation sequence:

$$\lambda_{k|i} = \alpha_0^i + \alpha_1^i \lambda_{k-1} + \cdots + \alpha_p^i \lambda_{k-p} + e_k \tag{5.2}$$

The model coefficients α_j^i, for $j = 0, \cdots, p$, depend on the system regime r_i for $i = 1, \cdots, m$ where m denotes the number of different possible regimes. The term e_k refers to the model residues and is the output of a random variable with some probability distribution function. Hence, $\lambda_{k|i}$ should be viewed as the value generated for observation λ_k when the process is at regime r_i. Notice that λ_k can be a vector with any arbitrary dimension. However, in this discussion, only univariate time-series are considered.

For any arbitrary regime, the current observation is computed as a weighted sum of past observations plus a term e_k, unpredictable, but whose probability density function is supposed to be known. Ahead, it is assumed that this random term follows a Gaussian distribution with zero mean and variance σ_e^2. Notice also that during this chapter, the models stability is always assumed. That is, the coefficients values α_j^i, for $j = 0, \cdots, p$ and $i = 1, \cdots, m$, are such that they ensure that the characteristic equation solutions are always inside the unity circle. This filter stability constraint is fundamental to ensure the model stationarity.

Defining $\boldsymbol{\alpha}_i = [\alpha_0^i \ \cdots \ \alpha_p^i]^T$ as the coefficients vector associated to regime r_i and $\boldsymbol{\lambda}_k = [1 \ \lambda_{k-1} \ \cdots \ \lambda_{k-p}]$ as the regressors vector, (5.2) can be formulated as,

$$\lambda_{k|i} = \boldsymbol{\lambda}_k \boldsymbol{\alpha}_i + e_k \tag{5.3}$$

It is important to notice that, even assuming exact knowledge about the model order, the coefficients values and the residues statistical properties, it is impossible to predict, with infinite accuracy, the value of λ_k for any regime r_i since e_k is, by definition, unpredictable. The only way is to estimate the expected value of $\lambda_{k|i}$. Defining $\hat{\lambda}_{k|i}$ as the expected value for $\lambda_{k|i}$, that is $\hat{\lambda}_{k|i} = \mathfrak{E}\{\lambda_{k|i}\}$, and taking into consideration both (5.2) and that the expected value of e_k is zero, the estimated value for $\lambda_{k|i}$ is obtained from:

$$\begin{aligned} \hat{\lambda}_{k|i} &= \mathfrak{E}\{\alpha_0^i + \alpha_1^i \lambda_{k-1} + \cdots + \alpha_p^i \lambda_{k-p} + e_k\} \\ &= \alpha_0^i + \alpha_1^i \hat{\lambda}_{k-1} + \cdots + \alpha_p^i \hat{\lambda}_{k-p} \end{aligned} \tag{5.4}$$

For the one step ahead prediction, where the observation λ_k is estimated by knowing all the past observation sequence, that is $\hat{\lambda}_{k-i} = \lambda_{k-i}$ for $i = 1, \cdots, p$, the previous expression can be presented alternatively as:

$$\begin{aligned} \hat{\lambda}_{k|i} &= \alpha_0^i + \alpha_1^i \lambda_{k-1} + \cdots + \alpha_p^i \lambda_{k-p} \\ &= \boldsymbol{\lambda}_k \boldsymbol{\alpha}_i \end{aligned} \tag{5.5}$$

In practice, the exact identification of all unknowns associated with this formulation is difficult, if not impossible. This is because all its degrees of freedom should be inferred, in most situations, based only on a limited set of observations generated by the process. All the knowledge concerning the number of different regimes, the regime switching mechanism, both the autoregressive parameters and model order, must be deduced only by looking at the data coming from the process output.

Consider *a priori* knowledge on both the number of regimes and the autoregressive model order. Suppose also that the regime switching mechanism follows the dynamics of a Markov chain. In this situation, for each distinct chain state corresponds a different process regime. For this reason, it only remains to compute both the Markov chain probabilities transition and the autoregressive models coefficients. All these values must be inferred by just considering a set of past process observations.

Figure 5.1 represents the ARMM structure associated to an arbitrary hidden state s_i. If the ARMM has m states then, for a complete representation of the ARMM, this image should be replicated m times.

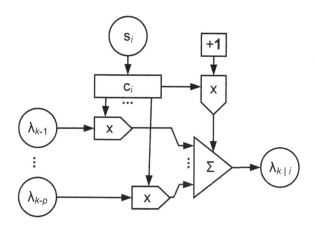

Figure 5.1 Partial structure of an autoregressive Markov model.

The value for λ_k, provided that the current active state is s_i, is obtained by an autoregressive filter whose coefficients are α_i. Hence, unlike the ordinary hidden Markov models, the observations depend on both the current active state and the previous observations. Is this dependence on past observations that leads to the temporal structure of ARMM.

Finally, the set of conditional observations $\lambda_{k|i}$, for $i = 1, \cdots, m$, will be used to compute the unconditional observation λ_k. In particular, the output λ_k is obtained through a weighted contribution of each of the conditional observations. This linear aggregation of conditional observations is illustrated in Figure 5.2. The set $\Lambda_k = \{\lambda_1, \cdots, \lambda_k\}$ consists on all the previous observations λ_i for $i = 1, \cdots, k$ and the weighting factors $P(q_k = s_i|\Lambda_k)$ refer to the probability of having s_i as the active state given the observations history Λ_k.

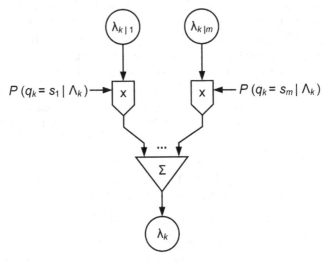

Figure 5.2 Computation of λ_k from the conditional observations $\lambda_{k|i}$ for $i = 1, \cdots, m$.

Before ending this section, a couple of important axioms are highlighted. Those axioms are fundamental to understand the operation of ARMM and are described below:

Axiom 1: The hidden states evolution follows the mechanism of a first order Markov chain. That is, the present active state only depends on the previous active state. This condition is expressed as:

$$P(q_k = s_j|q_{k-1} = s_i \wedge \cdots \wedge q_1 = s_l \wedge \Lambda_k) =$$
$$P(q_k = s_j|q_{k-1} = s_i) \tag{5.6}$$

Axiom 2: Let's assume that, for each operating regime, an observation λ_k has a p order autoregressive dependency on the previous observations and on the active chain states history. Then,

$$P(\lambda_k | q_k, \cdots, q_1 \wedge \lambda_{k-1}, \cdots, \lambda_1) =$$
$$P(\lambda_k | q_k, \cdots, q_{k-p} \wedge \lambda_{k-1}, \cdots, \lambda_{k-p}) \tag{5.7}$$

5.3 LIKELIHOOD AND PROBABILITY DENSITY FOR AR MODELS

As a preamble to the autoregressive Markov models theory, this section presents some important considerations regarding ordinary autoregressive models. In particular, their probability density function. Implicitly, this function is defined by taking into consideration both stationarity and knowledge on the statistical characteristics of the random disturbance e_k. In addition, during this section, the concept of likelihood of an AR model is introduced. This notion will assume that e_k is statistically described by a Gaussian distribution function.

5.3.1 AR MODEL PROBABILITY DENSITY FUNCTION

Suppose a p order autoregressive univariate model with the following structure:

$$\lambda_k = \alpha_0 + \alpha_1 \lambda_{k-1} + \cdots + \alpha_p \lambda_{k-p} + e_k \tag{5.8}$$

where e_k is assumed to be a value generated by a random process governed by a Gaussian distribution function with zero mean and variance σ_e^2.

Assuming that all the filter poles are inside the unity circle, which is a requirement for model stationarity, then the probability density function for the observations generating process is also Gaussian. Eventually with different mean and variance when compared to the ones of the disturbance component. Let's evaluate this statement by running a set of simulations. Those simulations are carried out assuming a third order autoregressive model with the following parametrization:

$$\lambda_k = \alpha_0 + \alpha_1 \lambda_{k-1} + \alpha_2 \lambda_{k-2} + \alpha_3 \lambda_{k-3} + e_k \tag{5.9}$$

By changing the values of the coefficients α_0 to α_3, the model poles can be moved arbitrarily inside the unity circle. Additionally, in this case, the random value e_k is generated by a Gaussian process with zero mean and unity variance.

Figure 5.3 shows the histograms for three different poles constellations. Each histogram was obtained by computing the absolute frequency over a total of 20,000 observations split into 100 bins. As can be seen, all the histogram shapes suggest a Gaussian process probability distribution function for λ.

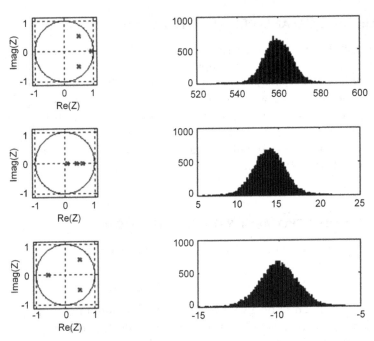

Figure 5.3 Observations histogram taken from a third order autoregressive model: on top $\lambda_k = 14 + 1.95\lambda_{k-1} - 1.45\lambda_{k-2} + 0.475\lambda_{k-3}$, at center $\lambda_k = 3 + 1.1\lambda_{k-1} - 0.34\lambda_{k-2} + 0.024\lambda_{k-3}$ and at bottom $\lambda_k = -8 + 0.4\lambda_{k-1} + 0.1\lambda_{k-2} - 0.3\lambda_{k-3}$.

The distribution center, assuming stationarity, can be easily obtained from the autoregressive model parameters. In order to do this, let's define μ_λ as the mean value of λ_k. By applying the expectation operator $\mathfrak{E}(\cdot)$ to (5.9) leads to:

$$\mathfrak{E}\{\lambda_k\} = \mathfrak{E}\{\alpha_0 + \alpha_1\lambda_{k-1} + \cdots + \alpha_p\lambda_{k-p} + e_k\} \qquad (5.10)$$

Taking into consideration that $\mathfrak{E}\{e_k\} = 0$ and that, if the system is stationary, $\mathfrak{E}\{\lambda_k\} = \mathfrak{E}\{\lambda_{k-1}\} = \cdots = \mathfrak{E}\{\lambda_{k-p}\} = \mu_\lambda$, then,

$$\mu_\lambda = \frac{\alpha_0}{1 - \sum_{i=1}^{p} \alpha_i} \tag{5.11}$$

Replacing α_0 in (5.8) by the previous equality, and after some algebraic manipulations, the following expression is obtained:

$$\lambda_k' = \sum_{i=1}^{p} \alpha_i \lambda_{k-i}' + e_k \tag{5.12}$$

where $\lambda_i' = \lambda_i - \mu_\lambda$.

The coefficients α_i can be computed using several methods. Among them, using the least squares method, through the Berg algorithm or even by solving the set of Yule-Walker equations.

Regarding this last method, the p Yule-Walker equations are obtained through the algebraic manipulations of (5.12). That is, multiplying both terms by λ_{k-l}', for $l = 1, \cdots, p$, and taking the expected values results in:

$$\mathfrak{E}\{\lambda_k' \lambda_{k-l}'\} = \sum_{i=1}^{p} \alpha_i \mathfrak{E}\{\lambda_{k-i}' \lambda_{k-l}'\} + \mathfrak{E}\{e_k \lambda_{k-l}'\} \tag{5.13}$$

Notice that $\mathfrak{E}\{e_k \lambda_{k-l}'\}$ is equal to σ_e^2 if $l = 0$ and zero otherwise. Indeed, this relation is straightforward to be demonstrated. Just take (5.8), multiply both terms by e_k and apply the expectation operator as shown below:

$$\begin{aligned} \mathfrak{E}\{e_k \lambda_k\} = &\alpha_0 \mathfrak{E}\{e_k\} + \alpha_1 \mathfrak{E}\{e_k \lambda_{k-1}\} + \cdots \\ &+ \alpha_p \mathfrak{E}\{e_k \lambda_{k-p}\} + \mathfrak{E}\{e_k e_k\} \end{aligned} \tag{5.14}$$

If e_k has zero mean value, then $\mathfrak{E}\{e_k e_k\}$ is its variance. Besides this, $\mathfrak{E}\{e_k \lambda_{k-i}\} = 0$ for $i \neq 0$, since $e_k \perp \lambda_{k-i}$. For $i = 0$ one has $\mathfrak{E}\{e_k \lambda_k\} = \sigma_e^2$. Additionally, defining $\mathfrak{E}\{\lambda_{k-i}' \lambda_{k-l}'\}$ as ρ_{l-i}, expression (5.13) can be represented as:

$$\rho_l = \sum_{i=1}^{p} \alpha_i \rho_{l-i} \tag{5.15}$$

for $l = 1, \cdots, p$.

This last expression is used to describe a set of p equations with p unknowns, whose matricial representation is:

$$
\begin{bmatrix} \rho_1 \\ \vdots \\ \rho_p \end{bmatrix} = \begin{bmatrix} \rho_0 & \rho_1 & \cdots & \rho_{p-1} \\ \vdots & \vdots & & \vdots \\ \rho_{p-1} & \rho_{p-2} & \cdots & \rho_0 \end{bmatrix} \begin{bmatrix} \alpha_1 \\ \alpha_2 \\ \vdots \\ \alpha_p \end{bmatrix}
\tag{5.16}
$$

This formulation can be compactly represented as $\rho = \mathbf{P}\alpha$ where $\mathbf{P}$ is a $p \times p$ Toeplitz matrix. Since $\mathbf{P}$ is a positive definite matrix, then the problem solution is obtained by solving,

$$
\alpha = \mathbf{P}^{-1}\rho
\tag{5.17}
$$

The coefficient α_0 from (5.8) can be calculated from:

$$
\alpha_0 = \mu_\lambda \left(1 - \sum_{i=1}^{p} c_i \right)
\tag{5.18}
$$

In practice, it is not possible to obtain the "true" values for the coefficients α since the true values of ρ are unknown. For this reason, assuming a set of N observations, the expected value of ρ is estimated using the sample mean:

$$
\hat{\rho}_{l-i} = \frac{1}{N-1} \sum_{k=1}^{N} \lambda'_{k-i} \lambda'_{k-l}
\tag{5.19}
$$

leading to,

$$
\hat{\alpha} = \hat{\mathbf{P}}^{-1}\hat{\rho}
\tag{5.20}
$$

The MATLAB® function presented in Listing 5.1 can be used to compute the value of $\hat{\rho}_{l-i}$ and the function in Listing 5.2 uses this last to estimate the autoregressive model parameters by the Yule-Walker method.

Listing 5.1 MATLAB® function to estimate ρ from a set of observations.

```
1 function r = rho(lambda,l,i)
2 r = sum(lambda(1+l:end-i).*lambda(1+i:end-l))/length(lambda);
```

Listing 5.2 MATLAB® function to estimating the autoregressive model coefficients by the Yule-Walker method.

```
1  function alfa = aryulewalker(lambda,p)
2  mu_lambda = mean(lambda);              % Compute mean
3  lambda_prime = lambda-mu_lambda;       % Lambda prime
4  Rox = zeros(1,p+1);                    % Initialize rho
5  for i=0:p,
6      Rox(1,i+1) = rho(lambda_prime,0,i);
7  end
8  Ro = Rox(2:end).';
9  P = toeplitz(Rox(1:end-1));            % Toeplitz matrix
10 alfa = P\Ro;                           % Prm. estimation
11 alfa = [mu_lambda*(1-sum(alfa));alfa].'; % + dc term
```

Assuming $l = 0$ in (5.13), then:

$$\mathfrak{E}\{\lambda'_k\lambda'_k\} = \sum_{i=1}^{p} \alpha_i \mathfrak{E}\{\lambda'_{k-i}\lambda'_k\} + \mathfrak{E}\{e_k\lambda'_k\} \tag{5.21}$$

Due to the fact that $\lambda'_k = \lambda_k - \mu_\lambda$, the mean value of λ'_k is always zero. Additionally, $\mathfrak{E}\{\lambda'_k\lambda'_k\} = \sigma^2_\lambda$ and the previous expression resumes to:

$$\sigma^2_\lambda = \sum_{i=1}^{p} \alpha_i \rho_i + \sigma^2_e \tag{5.22}$$

$$= \rho^T \alpha + \sigma^2_e$$

This final equality leads to the estimation of the autoregressive model variance based on both the model coefficients and the variance σ^2_e. Since the true values of α and σ^2_e are unknown, the model variance can only be estimated from:

$$\hat{\sigma}^2_\lambda = \hat{\rho}^T \hat{\alpha} + \hat{\sigma}^2_e \tag{5.23}$$

where $\hat{\sigma}^2_e$ can be inferred from the sample error variance. This error is computed by:

$$e_k = \lambda_k - \hat{\lambda}_k \tag{5.24}$$

for $\hat{\lambda}_k = \hat{\alpha}_0 + \sum_{i=1}^{p} \hat{\alpha}_i \lambda_{k-i}$.

In order to illustrate how the above expressions can be applied in a simulation context, consider a third order autoregressive stationary model given by:

$$\lambda_k = 2.3 + 1.95\lambda_{k-1} - 1.45\lambda_{k-2} + 0.475\lambda_{k-3} + e_k \tag{5.25}$$

where e_k represents values obtained by a random process with Gaussian probability distribution with zero mean and unity variance.

With this model, 20,000 observations are generated and this data was used to estimate the model parameters by the Yule-Walker method.

The parameters standard deviations are obtained by a bootstrap technique which consists on successive parameter estimations by adding the model generated observations to an error component which is obtained by randomization of the one step ahead error. The MATLAB® script described in Listing 5.3 executes the above described sequence of operations. The obtained estimated parameters, and respective standard deviations (represented within parentheses), lead to the following autoregressive model:

$$\hat{\lambda}_k = 2.326(0.1) + 1.932(0.007)\lambda_{k-1} - $$
$$1.419(0.012)\lambda_{k-2} + 0.462(0.065)\lambda_{k-3} \tag{5.26}$$

The obtained standard deviations indicate that, since the estimated parameters are close to the true parameters, all the parameters are sufficiently away from zero. By simple inspection it is possible to see that the closest is at around 23 standard deviations away from zero.

Listing 5.3 MATLAB® script for autoregressive model parameter and standard deviation estimation.

```
1  % Initialization
2  clear all;
3  clc;
4  close all;
5
6  N = 20000;                % Number of data samples
7  e = randn(1,N+200);       % Generate N->(0,1)
8  e = (e-mean(e)) / std(e); % Gaussian random var.
9  bootstrapsize = 200;      % Nbr. of bootstrap iter.
10
11 % Generate time series
12 lambda = zeros(1,N+200);  % Initialize lambda
13 alfa0=2.3;                % True parameters
14
15 % Poles location
16 alfa = -poly([0.95 0.5+0.5i 0.5-0.5i]);
17
18 % Generate time series
19 for k=4:N+200,
20     lambda(k) = alfa0+alfa(2)*lambda(k-1)+...
21                 alfa(3)*lambda(k-2)+...
22                 alfa(4)*lambda(k-3)+e(k);
23 end
24
25 % Remove initial transitory values
26 lambda(1:200)=[];
27 e(1:200)=[];              % (200 first samples)
```

```
28
29  % Parameters estimation by Yule-Walker method
30  alfae = aryulewalker(lambda,3);
31
32  % Parameters standard deviation estimation by bootstrap
33  estimation(1:3)=lambda(1:3);
34  for k=4:N,
35      % Observation estimation
36      estimation(k)=alfae(1)+alfae(2)*lambda(k-1)+...
37          alfae(3)*lambda(k-2)+alfae(4)*lambda(k-3);
38      % One step ahead prediction error
39      e(k)=lambda(k)-estimation(k);
40  end
41  par(1,:)=alfae;            % Estimated parameters
42  for l=2:bootstrapsize
43      inx=randperm(N);       % Randomize the error
44      e=e(inx);
45      % Compute new observations...
46      estimation(1:3)=lambda(1:3);
47      for k=4:length(lambda)
48          estimation(k)=alfae(1) +...
49          alfae(2)*estimation(k-1) +...
50          alfae(3)*estimation(k-2) + ...
51          alfae(4)*estimation(k-3) + e(k);
52      end
53      % Estimate new coefficients
54      par(1,:) = aryulewalker(estimation,3);
55  end
56
57  % Compute the parameters standard deviation
58  std(par)
```

Taking into consideration the coefficients of (5.25), and applying (5.18), the obtained time series mean value was 92 and the sampling mean obtained from the observations was 91.996. Additionally, the autoregressive variance value, after applying (5.23), was 43.456. A value coherent with 43.48 was obtained from the computed observations variance.

5.3.2 AUTOREGRESSIVE MODEL LIKELIHOOD

Let's consider a sequence of N observations $\lambda_1, \cdots, \lambda_N$ where, λ_1 was the first observation to take place and λ_N was the last. Assuming a p order stationary autoregressive model, as the one described at (5.8), the objective is to establish the likelihood of this model based on the observed sequence. That is, to determine the probability of this set of observations that have been generated by the model (5.8) knowing, *a priori*, the coefficients values and the statistical behaviour for e_k. This condition is denoted by $P(\lambda_N, \cdots, \lambda_p, \cdots, \lambda_1 | \boldsymbol{\alpha} \wedge \sigma_e^2)$ where $\boldsymbol{\alpha} = \{\alpha_0, \cdots, \alpha_p\}$ and σ_e^2 is the e_k variance. In order to simplify the notation, consider that the dependence on α and σ_e^2 is implicit. That is, the above probability can be represented simply by $P(\lambda_N, \cdots, \lambda_p, \cdots, \lambda_1)$.

From Bayes theorem, the model likelihood can then be computed as:

$$
\begin{aligned}
P(\lambda_N, \cdots, \lambda_p, \cdots, \lambda_1) = \\
P(\lambda_p, \cdots, \lambda_1) \prod_{i=p+1}^{N} P(\lambda_i | \lambda_{i-1}, \cdots, \lambda_p, \cdots, \lambda_1)
\end{aligned} \tag{5.27}
$$

Taking into consideration the second axiom provided in Section 5.2, for a p order autoregressive model, the observation λ_i only depends on $\lambda_{i-1}, \cdots, \lambda_{i-p}$. That is,

$$
P(\lambda_i | \lambda_{i-1}, \cdots, \lambda_p, \cdots, \lambda_1) = P(\lambda_i | \lambda_{i-1}, \cdots, \lambda_{i-p}) \tag{5.28}
$$

as long as $i - p > 0$. For this reason, (5.27) can be rewritten as:

$$
\begin{aligned}
P(\lambda_N, \cdots, \lambda_p, \cdots, \lambda_1) = \\
P(\lambda_p, \cdots, \lambda_1) \prod_{i=p+1}^{N} P(\lambda_i | \lambda_{i-1}, \cdots, \lambda_{i-p})
\end{aligned} \tag{5.29}
$$

This means that, for the likelihood value of an observations set using an autoregressive model with parameters $\mathbf{c}$, it is necessary to obtain the marginal and conditional probabilities.

Starting from the marginal probability, the observation sequence $\{\lambda_p, \cdots, \lambda_1\}$ can be viewed as a p dimensional vector denoted by $\boldsymbol{\lambda}$. Additionally, assuming a Gaussian distribution for e_k, then λ_k has also a Gaussian distribution. For this reason, the marginal probability is $P(\lambda_p, \cdots, \lambda_1) = \mathcal{G}(\boldsymbol{\lambda}, \boldsymbol{\mu}_\lambda, \boldsymbol{\Sigma}_\lambda)$ where $\mathcal{G}(\boldsymbol{\lambda}, \boldsymbol{\mu}_\lambda, \boldsymbol{\Sigma}_\lambda)$ is the following multivariate Gaussian function:

$$
\mathcal{G}(\boldsymbol{\lambda}, \boldsymbol{\mu}_\lambda, \boldsymbol{\Sigma}_\lambda) = \frac{\exp\left(-\frac{1}{2}(\boldsymbol{\lambda} - \boldsymbol{\mu}_\lambda)^T \boldsymbol{\Sigma}_\lambda^{-1}(\boldsymbol{\lambda} - \boldsymbol{\mu}_\lambda)\right)}{\sqrt{\det(2\pi\boldsymbol{\Sigma}_\lambda)}} \tag{5.30}
$$

and the vector $\boldsymbol{\mu}_\lambda$ is equal to $\mu_\lambda \cdot \mathbf{1}$ for μ_λ computed from (5.11). Moreover, the $p \times p$ covariance matrix $\boldsymbol{\Sigma}_\lambda$ is:

$$
\boldsymbol{\Sigma}_\lambda = \begin{bmatrix} \sigma_p^2 & \sigma_p\sigma_{p-1} & \cdots & \sigma_p\sigma_1 \\ \sigma_{p-1}\sigma_p & \sigma_{p-1}^2 & & \sigma_{p-1}\sigma_1 \\ \vdots & \vdots & \ddots & \vdots \\ \sigma_1\sigma_p & \sigma_1\sigma_{p-1} & \cdots & \sigma_1^2 \end{bmatrix} \tag{5.31}
$$

Since the process is stationary, all the elements on the main diagonal are equal. That is, $\sigma_p^2 = \sigma_{p-1}^2 = \cdots = \sigma_1^2 = \sigma_\lambda^2$.

Regarding the conditional probability $P(\lambda_i | \lambda_{i-1}, \cdots, \lambda_{i-p})$, first the knowledge of $\lambda_{i-1}, \cdots, \lambda_{i-p}$ is assumed. Hence, those observations are treated as constants in the autoregressive model. For this reason, the new value for μ_λ is:

$$\mathfrak{E}\{\lambda_i\} = \mathfrak{E}\{\alpha_0 + \alpha_1 \lambda_{i-1} + \cdots + \alpha_p \lambda_{i-p} + e_i\}$$
$$= \alpha_0 + \alpha_1 \lambda_{i-1} + \cdots + \alpha_p \lambda_{i-p} = \mu_i \tag{5.32}$$

which changes with the index i.

On the other hand, the new value for the variance σ_λ^2 is:

$$\mathfrak{E}\{(\lambda_k - \mu_k)^2\} = \mathfrak{E}\{e_k^2\} = \sigma_e^2 \tag{5.33}$$

In this framework, and since the value of μ_i corresponds to the estimated value $\hat{\lambda}_i$, $\lambda_i - \mu_i = \lambda_i - \hat{\lambda}_i = e_i$ this leads to the following conditional probability expression:

$$P(\lambda_i | \lambda_{i-1}, \cdots, \lambda_{i-p}) = \frac{1}{\sqrt{2\pi\sigma_e^2}} \exp\left(-\frac{e_i^2}{2\sigma_e^2}\right) \tag{5.34}$$

It can be seen that the likelihood of an autoregressive model is the product of Gaussian functions. This product can be translated into sums using a logarithm transformation. That is,

$$\mathcal{L}^e = \ln\{P(\lambda_p, \cdots, \lambda_1)\} + \sum_{i=p+1}^{N} \ln\{P(\lambda_i | \lambda_{i-1}, \cdots, \lambda_{i-p})\} \tag{5.35}$$

At this point, it is possible to compute the likelihood function of a regular autoregressive parametric model. In the following section, this knowledge will be used to derive an expression for the likelihood of an ARMM.

5.4   LIKELIHOOD OF AN ARMM

In an ARMM with m hidden states, a total of m distinct linear autoregressive models can be effectively found. However, at a given time instant, only one of them is active. For this reason, many of the considerations presented during Section 5.3, can be used and extrapolated to the present discussion.

First let's assume, without loss of generality, that all the m autoregressive models have the same order denoted, hereafter, by p. If the present active Markov hidden state is s_i, for $i = 1, \cdots, m$, then the current autoregressive model output is:

$$\lambda_{k|i} = \alpha_0^i + \alpha_1^i \lambda_{k-1} + \cdots + \alpha_p^i \lambda_{k-p} \tag{5.36}$$

From the previous section, and assuming a random Gaussian type disturbance with zero mean and variance σ_e^2, the autoregressive model likelihood is given by:

$$P(\lambda_k | \Lambda_{k-1}) = \frac{1}{\sqrt{2\pi\sigma_e^2}} \exp\left(-\frac{e_k^2}{2\sigma_e^2}\right) \tag{5.37}$$

However, within an ARMM framework, the observation λ_k can be generated by any of the available m models. That is,

$$P(\lambda_k | \Lambda_{k-1}) = \sum_{i=1}^{m} P(\lambda_k | q_k = s_i \wedge \Lambda_{k-1}) P(q_k = s_i | \Lambda_{k-1}) \tag{5.38}$$

where

$$P(\lambda_k | q_k = s_i \wedge \Lambda_{k-1}) = \frac{\exp\left(-\frac{1}{2\sigma_e^2}(e_k^i)^2\right)}{\sqrt{2\pi\sigma_e^2}} \tag{5.39}$$

and

$$e_k^i = \lambda_k - \alpha_0^i - \alpha_1^i \lambda_{k-1} - \cdots - \alpha_p^i \lambda_{k-p} \tag{5.40}$$

For a sequence of N observations Λ_N, the likelihood can be computed in the following way:

$$\begin{aligned} P(\Lambda_N) &= P(\lambda_N \wedge \Lambda_{N-1}) \\ &= P(\lambda_N | \Lambda_{N-1}) \cdot P(\Lambda_{N-1}) \end{aligned} \tag{5.41}$$

Applying the same idea to $P(\Lambda_{N-1})$ leads to:

$$P(\Lambda_{N-1}) = P(\lambda_{N-1} | \Lambda_{N-2}) \cdot P(\Lambda_{N-2}) \tag{5.42}$$

This approach can be repeated up to Λ_p. At that moment, the following equality is verified:

$$P(\Lambda_{p+1}) = P(\lambda_{p+1} | \Lambda_p) \cdot P(\Lambda_p) \tag{5.43}$$

Finally, $P(\Lambda_N)$ takes the form:

$$P(\Lambda_N) = P(\Lambda_p) \prod_{k=p+1}^{N} P(\lambda_k|\Lambda_{k-1}) \tag{5.44}$$

where $P(\lambda_k|\Lambda_{k-1})$ was already defined at (5.38).

Usually, it is computationally more efficient to replace the previous function, by the first p observations conditional likelihood [12]. This modification can be achieved by dividing both members of (5.44) by $P(\Lambda_p)$ leading to:

$$\frac{P(\Lambda_N)}{P(\Lambda_p)} = P(\Lambda_{N\to p+1}|\Lambda_p) = \prod_{k=p+1}^{N} P(\lambda_k|\Lambda_{k-1}) \tag{5.45}$$

where $\Lambda_{N\to p+1}$ denotes the set of observations taken from time $p+1$ to N. With this simple modification, it is no longer required to take into consideration the first p observations. For this reason, denoting by $\mathcal{L}_{N|p}$ the likelihood of a given autoregressive Markov model, and using (5.45), the following expression is presented,

$$\mathcal{L}_{N|p} = \prod_{k=p+1}^{N} P(\lambda_k|\Lambda_{k-1}) \tag{5.46}$$

or using the log-likelihood formulation,

$$\mathcal{L}^e_{N|p} = \sum_{k=p+1}^{N} \ln\{P(\lambda_k|\Lambda_{k-1})\} \tag{5.47}$$

5.5 ARMM PARAMETERS ESTIMATIONS

Now that the likelihood function of an ARMM is established, it is time to address the question concerning the parameters estimation. The autoregressive Markov model parameter estimation procedure consists on finding the vector $\mathbf{c}$ and the transition probabilities matrix $\mathbf{A}$ such as to maximize the conditional likelihood given by (5.46). In order to do this, it is necessary to find a way to efficiently compute $P(q_k = s_i|\Lambda_{k-1})$. Observe that trying to express it in something like $P(q_{k-1} = s_i \wedge \Lambda_{k-1})$ to then use the forward algorithm is out of the question. This is because the forward algorithm can only be used assuming conditional independence between observations and that is not the case when dealing with ARMM. Thus, it is necessary to establish an alternative way to calculate it. Since $P(\Lambda_k)$ is equal to $P(\lambda_k \wedge \Lambda_{k-1})$, the probability $P(q_k = s_i | \Lambda_k)$ can be rewritten as:

$$P(q_k = s_i | \Lambda_k) = P(q_k = s_i | \lambda_k \wedge \Lambda_{k-1}) \tag{5.48}$$

After applying Bayes theorem, followed by the chain rule, the following expression is obtained:

$$P(q_k = s_i | \Lambda_k) = \frac{P(\lambda_k | q_k = s_i \wedge \Lambda_{k-1}) P(q_k = s_i | \Lambda_{k-1}) P(\Lambda_{k-1})}{P(\lambda_k \wedge \Lambda_{k-1})} \tag{5.49}$$

which, after dividing both the numerator and dominator by $P(\Lambda_{k-1})$ and applying again the Bayes rule, leads to:

$$P(q_k = s_i | \Lambda_k) = \frac{P(\lambda_k | q_k = s_i \wedge \Lambda_{k-1}) P(q_k = s_i | \Lambda_{k-1})}{P(\lambda_k | \Lambda_{k-1})} \tag{5.50}$$

Finally, resorting to expression (5.38), then,

$$P(q_k = s_i | \Lambda_k) = \frac{P(\lambda_k | q_k = s_i \wedge \Lambda_{k-1}) P(q_k = s_i | \Lambda_{k-1})}{\sum\limits_{j=1}^{m} P(\lambda_k | q_k = s_j \wedge \Lambda_{k-1}) P(q_k = s_j | \Lambda_{k-1})} \tag{5.51}$$

Due to the fact that,

$$
\begin{aligned}
P(q_k = s_i | \Lambda_{k-1}) &= \\
&\sum_{j=1}^{m} P(q_k = s_i | q_{k-1} = s_j) P(q_{k-1} = s_j | \Lambda_{k-1}) \\
&= \sum_{j=1}^{m} a_{ji} P(q_{k-1} = s_j | \Lambda_{k-1})
\end{aligned}
\tag{5.52}
$$

and since the Markov model is homogeneous in time, then:

$$P(q_{k+1} = s_i | \Lambda_k) = \sum_{j=1}^{m} a_{ji} P(q_k = s_j | \Lambda_k) \tag{5.53}$$

Taking into consideration (5.51) and (5.53), a recurrence formula for computing $P(q_k = s_i | \Lambda_{k-1})$ is then defined. The iterative procedure is illustrated by focusing on the computation of $P(q_2 = s_i | \Lambda_1)$. Let's start by assuming that,

$$P(q_1 = s_i | \Lambda_0) = P(q_1 = s_i) = \alpha_i \tag{5.54}$$

Then, the computation of $P(q_1 = s_i | \Lambda_1)$ is obtained from:

$$
\begin{aligned}
P(q_1 = s_i | \Lambda_1) &= \frac{P(\lambda_1 | q_1 = s_i \wedge \Lambda_0) P(q_1 = s_i | \Lambda_0)}{\sum_{j=1}^{m} P(\lambda_1 | q_1 = s_j \wedge \Lambda_0) P(q_1 = s_j | \Lambda_0)} \\
&= \frac{P(\lambda_1 | q_1 = s_i \wedge \Lambda_0) \cdot \alpha_i}{\sum_{j=1}^{m} P(\lambda_1 | q_1 = s_j \wedge \Lambda_0) \cdot \alpha_j}
\end{aligned}
\tag{5.55}
$$

Using the above result, and after replacing at (5.53), leads to:

$$P(q_2 = s_i | \Lambda_1) = \sum_{j=1}^{m} a_{ji} \cdot P(q_1 = s_j | \Lambda_1) \tag{5.56}$$

The above procedure's recursive approach, which will be designated by Hamilton algorithm, is coded for MATLAB® and presented in Listing 5.4. Notice that the reference within the code to the variables P0, P1 and P2 denotes the probabilities $P(\lambda_k | q_k \wedge \Lambda_{k-1})$, $P(q_k = s_i | \Lambda_{k-1})$ and $P(q_k = s_i | \Lambda_k)$, respectively.

Listing 5.4 MATLAB® code for Hamilton's algorithm computation.

```
1  function [P1,P2,lik]=hamilton_algorithm(A,c,Lambda,alfa,sigma2)
2  % Parameters initialization
3  Lambda = Lambda(:);    % Lambda is a column vector
4  [n,m] = size(alfa);    % order (p) and states (m)
5  p = n-1;               % p = nbr coeff. - 1
6  N = length(Lambda);    % Number of observations.
7  lik = 0;               % Model likelihood
8  P0 = zeros(m,N);       % P(\lambda_k/q_k ^ \Lambda_{k-1})
9  P1 = zeros(m,N);       % P(q_k=s_i/\Lambda_{k-1})
10 P2 = zeros(m,N);       % P(q_k=s_i/\Lambda_k)
11 % Impose P(q_k=s_i/\Lambda_{k-1})=c_i for k = p+1
12 k = p + 1;
13 for i=1:m,
14     P1(i,k)=c(i);
15 end
16 for k=p+1:N-1,
17     % Compute P(q_k=s_i/\Lambda_{k}) and
18     % P(\lambda_k/q_k \wedge \Lambda_{k-1})
19     for i=1:m,
20         h=[1;Lambda(k-1:-1:k-p)];
```

```
21        P0(i,k)=(1/(sqrt(2*pi*sigma2)))*...
22      exp(-((Lambda(k)-h'*alfa(:,i)).^2)/(2*sigma2));
23        P2(i,k)=P0(i,k)*P1(i,k);
24    end
25    denominator=sum(P2(:,k));
26    lik=lik+log(denominator);
27    for i=1:m,
28        P2(i,k)=P2(i,k)/denominator;
29    end
30    % Compute  P(q_k=s_i/\Lambda_{k-1})
31    for i=1:m,
32        P1(i,k+1)=0;
33        for j=1:m,
34            P1(i,k+1)=P1(i,k+1)+A(j,i)*P2(j,k);
35        end
36    end
37 end
38 % Compute   P(q_k=s_i/\Lambda_{k})  for  k = N
39 k=N;
40 for i=1:m,
41    h=[1;Lambda(k-1:-1:k-p)];
42    P0(i,k)=(1/(sqrt(2*pi*sigma2)))*...
43    exp(-((Lambda(k)-h'*alfa(:,i)).^2)/(2*sigma2));
44    P2(i,k)=P0(i,k)*P1(i,k);
45 end
46 denominator=sum(P2(:,k));
47 lik=lik+log(denominator);
48 for i=1:m,
49    P2(i,k)=P2(i,k)/denominator;
50 end
```

5.5.1 PARAMETERS ESTIMATION

One of the major tasks during ARMM synthesis is to estimate the unknown model parameters. That is, the $m \times (p+1)$ autoregressive models coefficients, the state transition matrix $\mathbf{A}$, the priors $\mathbf{c}$ and the residues variance σ_e^2. As already highlighted, this estimation must be carried out using only the information conveyed in the observations sequence which is measured at the output of some unknown stochastic process.

In order to perform the parameter estimation task, let's start by assuming an ARMM with m states. Each one of these states controls the parameters of one of the AR models. Consider that the model associated to the present state q_k has the following structure:

$$\begin{aligned} \lambda_k &= \alpha_0^{q_k} + \alpha_1^{q_k}\lambda_{k-1} + \cdots + \alpha_p^{q_k}\lambda_{k-p} + e_k \\ &= \boldsymbol{\lambda}_k \boldsymbol{\alpha}^{q_k} + e_k \end{aligned} \tag{5.57}$$

where $\boldsymbol{\lambda}_k = [1 \quad \lambda_{k-1} \quad \cdots \quad \lambda_{k-p}]$ and $\boldsymbol{\alpha}^{q_k} = [\alpha_0^{q_k} \quad \alpha_1^{q_k} \quad \cdots \quad \alpha_p^{q_k}]^T$. The parameters $\boldsymbol{\alpha}^{q_k}$ denote the autoregressive model coefficients associated to an arbitrary active state $q_k = s_i$ for $i = 1, \cdots m$ at present time k. Once again, it is assumed that the value of e_k is generated by a Gaussian distribution random process with zero mean and variance σ_e^2. For this model, the probability of observing λ_k, provided the knowledge of all past observations and states (including the present), is:

$$P(\lambda_k|\Lambda_{k-1} \wedge Q_k) = \frac{\exp\left(-\frac{(\lambda_k - \lambda_k \alpha^{q_k})^2}{2\sigma^2}\right)}{\sqrt{2\pi\sigma^2}} \qquad (5.58)$$

where $Q_k = \{q_1, \cdots, q_k\}$ and, due to stationarity, σ^2 is the model variance which is assumed constant.

For a particular state, for example s_i where i is an integer between 1 and m, the associated autoregressive model parameters $\boldsymbol{\alpha}^{s_i}$ are determined iteratively. The iterative equation is obtained from the expected value of the conditional log-likelihood given by $\log\left[P(\Lambda_{p+1 \to N}|\Lambda_p)\right]$ and represented as [11, 12]:

$$\mathbf{c}^{s_i} = \left[\sum_{k=p+1}^{N} P(q_k = s_i|\Lambda_N)\boldsymbol{\lambda}_k\boldsymbol{\lambda}_k^T\right]^{-1} \left[\sum_{k=p+1}^{N} P(q_k = s_i|\Lambda_N)\boldsymbol{\lambda}_k^T\lambda_k\right] \qquad (5.59)$$

which is nothing more than the mathematical formulation of the parameter estimation method from the weighted least squares method. In this case, the weighting factors are the states activation probabilities.

In the weighted least squares estimation method, different levels of significance are assigned to the observations. This differentiation is carried out by defining a set $\{w_1, \cdots, w_N\}$ of different weights. In this context, the least squares cost function is:

$$J(\boldsymbol{\alpha}) = \sum_{i=p+1}^{N} w_i (\lambda_i - \lambda_i\boldsymbol{\alpha})^2 \qquad (5.60)$$

or alternatively, in matricial format, as:

$$J(\boldsymbol{\alpha}) = \mathbf{E}^T\mathbf{W}\mathbf{E} \qquad (5.61)$$

where $\mathbf{W}$ is a $(N - p) \times (N - p)$ diagonal matrix whose elements are the weight values associated to each observation and $\mathbf{E} = \boldsymbol{\Lambda} - \mathbf{H}\boldsymbol{\alpha}$ for,

$$\mathbf{H} = \begin{bmatrix} \boldsymbol{\lambda}_{p+1} & \cdots & \boldsymbol{\lambda}_N \end{bmatrix}^T \qquad (5.62)$$

a $(N - p) \times (p + 1)$ regression vectors matrix and $\mathbf{\Lambda} = [\lambda_{p+1} \cdots \lambda_N]^T$ a $(N - p)$ observations column vector.

The model coefficients $\boldsymbol{\alpha}$ are computed, in closed form, by an expression obtained after finding the critical points of $J(\boldsymbol{\alpha})$. That is, solving for,

$$\frac{\partial J(\boldsymbol{\alpha})}{\partial \boldsymbol{\alpha}} = 0$$

which yields:

$$\mathbf{c} = \left(\mathbf{H}^T \mathbf{W} \mathbf{H}\right)^{-1} \mathbf{H}^T \mathbf{W} \mathbf{\Lambda} \qquad (5.63)$$

Let the autoregressive model coefficients, associated to state $q_k = s_i$, be represented by $\boldsymbol{\alpha}^{s_i} = \left[\alpha_0^{s_i} \ \alpha_1^{s_i} \ \cdots \ \alpha_p^{s_i}\right]^T$ and the weight matrix, defined as function of the active state s_i, as:

$$\mathbf{W}_{s_i} = \left(\left[P(q_{p+1} = s_i|\Lambda_N) \ \cdots \ P(q_N = s_i|\Lambda_N)\right]\right) \qquad (5.64)$$

where $\Lambda_N = \{\lambda_1, \cdots, \lambda_N\}$. Then, (5.59) can be translated into matricial form as:

$$\boldsymbol{\alpha}_{s_i} = \left(\mathbf{H}^T \mathbf{W}_{s_i} \mathbf{H}\right)^{-1} \mathbf{H}^T \mathbf{W}_{s_i} \mathbf{\Lambda} \qquad (5.65)$$

Notice the similarity between this last expression and the one in (5.63). Moreover, and in order to complete the computation method for (5.59), it is necessary to unveil a way to obtain $P(q_k = s_i|\Lambda_N)$. This probability can be computed iteratively by a procedure known as Kim's algorithm [12]. In Kim's algorithm the value of $P(q_k = s_i|\Lambda_N)$ is obtained according to:

$$P(q_k = s_i|\Lambda_N) = P(q_k = s_i|\Lambda_k) \sum_{j=1}^{m} a_{ij} \frac{P(q_{k+1} = s_j|\Lambda_N)}{P(q_{k+1} = s_j|\Lambda_k)} \qquad (5.66)$$

Taking into consideration that their values are computed in reverse order, starting from $k = N - 1$ down to $k = 1$, the method begins with $P(q_N = s_i|\Lambda_N)$, whose value is obtained by replacing k by N at (5.51). That is,

$$P(q_N = s_i|\Lambda_N) = \frac{P(q_N = s_i|\Lambda_{N-1})P(\lambda_N|q_N = s_i \wedge \Lambda_{N-1})}{\sum_{j=1}^{m} P(q_N = s_j|\Lambda_{N-1})P(\lambda_N|q_N = s_j \wedge \Lambda_{N-1})} \qquad (5.67)$$

Kim's algorithm is presented as a MATLAB® function in Listing 5.5. In this function, the pair of variables P1 and P2 refers to the homonym variables at Listing 5.4 and variable P3 regards the probability $P(q_k = s_i|\Lambda_N)$.

The model variance σ^2 value is estimated as a weighted average of the residue's variance taken from the m different autoregressive models. This can be expressed as:

$$\sigma^2 = \frac{1}{N-p} \sum_{k=p+1}^{N} \sum_{i=1}^{m} P(q_k = s_i | \Lambda_N) \left(\lambda_k - \boldsymbol{\lambda}_k \boldsymbol{\alpha}^{q_k} \right)^2 \qquad (5.68)$$

Please observe the explicit use of the previously calculated value of $P(q_k = s_i | \Lambda_N)$.

The autoregressive models parameters estimation, using (5.65) and assuming the model variance estimated by (4.52), can be performed by the MATLAB® function presented at Listing 5.7.

In this function, the variable P3 regards the probability $P(q_k = s_i | \Lambda_N)$ obtained after Kim's algorithm. Listing 5.6 concerns a MATLAB® script that can be used to test both the Hamilton and Kim's algorithms.

A time-series is first generated and the AR hidden Markov model parameters are randomly initialized. Then, the above described algorithms are executed in order to obtain the probabilities.

Listing 5.5 MATLAB® version for Kim's algorithm.

```
1  function P3 = kim_algorithm(A,P1,P2,p)
2
3  % Parameters initialization
4  [m,N]=size(P1);
5  P3=zeros(m,N);
6  for i=1:m,
7      P3(i,N)=P2(i,N);
8  end
9
10 % Recursion
11 for k=N-1:-1:p+1
12     for i=1:m,
13         second_term=0;
14         for j=1:m
15             second_term=second_term+A(i,j)*P3(j,k+1)/P1(j,k+1);
16         end
17         P3(i,k)=P2(i,k)*second_term;
18     end
19 end
```

Listing 5.6 MATLAB® script to test both the Hamilton and Kim's algorithms.

```
1  % Generate the time-series
2  clear all;
3  clc;
4  y(1)=0;y(2)= 0.06;y(3)=0.2;y(4)=0.4;y(5)=0.5;
5  for k=5:200,
6      y(k)=2.7*y(k-1)-3.1*y(k-2)+1.8*y(k-3)-0.48*y(k-4)+0.01*randn;
7  end
8
9  plot(y);
10 xlabel('Time');
11 ylabel('Amplitude');
12 grid on
13
14 % AR Markov model parameters
15 m=2;  % number of hidden states
16 p=2;  % AR model order
17
18 % Initial probabilities vector
19 c=rand(m,1);
20 c=c/sum(c);
21
22 % States transition matrix
23 A=rand(m,m);
24 A=A./(sum(A,1)'*ones(1,m));
25
26 % Initialize Sigma
27 sigma=1;
28
29 % Initialize model parameters ((p+1) x m) in a way as to
30 % force model stability.
31 for i=1:m,
32     poles=rand(1,p);
33     eqx=poly(poles);
34     alfa(:,i)=[rand -eqx(2:end)].';
35 end
36
37 % Run the Hamilton algorithm
38 [P1,P2]=hamilton_algorithm(A,c,y,alfa,sigma);
39
40 % Run the Kim algorithm
41 P3=kim_algorithm(A,P1,P2,p);
```

Listing 5.7 MATLAB® code for computing variance and autoregressive model coefficients.

```
1  function [alfa,sigma2]=markov_ar_coef(p,Lambda,P3)
2
3  % Parameters initialization
4  Lambda=Lambda(:);    % Lambda is a column vector
5  [m,N]=size(P3);      % States (m) and observ. (N)
6  alfa=zeros(p+1,m);   % Pre-allocate space
7  Sigma=zeros(1,m);    % Pre-allocate space
8  H=zeros(N-p,p+1);    % Pre-allocate space
9  H(:,1)=1;
10
11 % Regressors Matrix
```

```
12 for i=1:p,
13      H(:,i+1)=Lambda(p-i+1:end-i);
14 end
15 Y=Lambda(p+1:end);
16
17 % Weighted least squares parameters estimation
18 for i=1:m,
19      X=P3(i,p+1:end);
20      W=diag(X);
21      alfa(:,i)=(H'*W*H)\ (H'*W*Lambda(p+1:end));
22 end
23
24 % Variance estimation
25 for i=1:m
26      E=Y-H*alfa(:,i);
27      X=P3(i,p+1:end);
28      W=diag(X);
29      Sigma(i)=E'*W*E;
30 end
31
32 sigma2=sum(Sigma)/(N-p);
```

Finally, in order to complete the ARMM parameter estimation procedure, it is necessary to obtain the state transition probabilities and the value for the priors. In [11], the probability transition from state s_i to state s_j is obtained iteratively using:

$$\hat{a}_{ij} = \frac{\sum_{k=p+1}^{N} P(q_k = s_j \wedge q_{k-1} = s_i | \Lambda_N)}{\sum_{k=p+1}^{N} P(q_{k-1} = s_i | \Lambda_N)} \tag{5.69}$$

where the marginal probability at the denominator requires also the values obtained from Kim's algorithm.

Furthermore, the joint probability at the numerator must be deduced taking into consideration only the marginal probabilities values. By applying Bayes theorem, the joint probability can be rewritten as:

$$P(q_k = s_j \wedge q_{k-1} = s_i | \Lambda_N) = P(q_k = s_j | q_{k-1} = s_i) P(q_{k-1} = s_i | \Lambda_N)$$
$$= a_{ij} P(q_{k-1} = s_i | \Lambda_N) \tag{5.70}$$

and, after replacing it into (5.69), results in:

$$\hat{a}_{ij} = \frac{\sum_{k=p+1}^{N} a_{ij} \cdot P(q_{k-1} = s_i | \Lambda_N)}{\sum_{k=p+1}^{N} P(q_{k-1} = s_i | \Lambda_N)} \tag{5.71}$$

Finally, the values for the priors are estimated using:

$$\hat{c}_i = P(q_1 = s_i | \Lambda_N) \tag{5.72}$$

for $i = 1, \cdots, m$.

Both the probabilities transition matrix $\mathbf{A}$ and priors vector $\mathbf{c}$ can be computed in MATLAB® by using the code presented at Listing 5.8. Once again the variable P3 regards the probabilities obtained after Kim's algorithm.

To conclude this section, the ARMM iterative parameter estimation procedure is summarized in the following sequence of six steps. It is important to highlight that the presented sequence of equations assume order p for all the autoregressive models and a total of m hidden states. Moreover, it is assumed an univariate time series composed by a total of N observations and the index l is the iteration counter that takes integer values from 1 to an user defined maximum.

1st step: Compute $P(q_k = s_j | \Lambda_k)$ for $j = 1, \cdots, m$ and $k = p + 1, \cdots, N$ recursively using both (5.51) and (5.53). The algorithm initialization is done with:

$$P(q_k = s_j | \Lambda_{k-1}) = c_j \tag{5.73}$$

for $k = p + 1$.

2nd step: Using Kim's algorithm described at (5.66), recursively compute $P(q_k = s_j | \Lambda_N)$ for $j = 1, \cdots, m$ and $k = N - 1, \cdots, p + 1$. The process is initialized with the previous step value for $P(q_N | \Lambda_N)$.

3rd step: Compute the states transition probability matrix using:

$$\hat{a}_{ij}^{l+1} = \frac{\sum_{k=p+1}^{N} \hat{a}_{ij}^l P(q_{k-1} = s_i | \Lambda_N)}{\sum_{k=p+1}^{N} P(q_{k-1} = s_i | \Lambda_N)} \tag{5.74}$$

where the probabilities values involved are the ones obtained in the second step.

4th step: Calculate the prior associated to state i from:

$$\hat{c}_i^{l+1} = P(q_k = s_i | \Lambda_N) \tag{5.75}$$

for $k = p + 1$.

5th step Estimate the m autoregressive model coefficients from:

$$\hat{\boldsymbol{\alpha}}_{s_i}^{l+1} = (\mathbf{H}^T \mathbf{W}_{s_i} \mathbf{H})^{-1} (\mathbf{H}^T \mathbf{W}_{s_i} \boldsymbol{\Lambda}) \tag{5.76}$$

6th step Compute the new value of σ^2 from:

$$\hat{\sigma}_{l+1}^2 = \frac{1}{N-p}\sum_{i=1}^{m}\mathbf{E}_i^T\mathbf{W}_i\mathbf{E}_i \qquad (5.77)$$

where,

$$\mathbf{E}_i = \boldsymbol{\Lambda} - \mathbf{H}\hat{\alpha}_i \qquad (5.78)$$

Stoping criteria: The iterative procedure can be stopped when the maximum number of iterations imposed by the user is reached or, for example, when the maximum difference value between successive solutions is below a given threshold (for example 10^{-8})[11].

All the above steps are implemented within the MATLAB® function presented at Listing 5.9.

Listing 5.8 MATLAB® code for computing the transition probability matrix **A** and the vector **c**.

```
1  function [A_new,c]=markov_ar_matx(p,A,P3)
2
3  % Parameters initialization
4  [m,N]=size(P3);   % states (m) and observations(N)
5  A_new=zeros(m,m); % Pre-allocate space
6
7  % Estimation of the new probabilities transition matrix
8  for i=1:m
9      for j=1:m
10         numerator=0;
11         for k=p+1:N,
12             numerator=numerator+A(i,j)*P3(i,k-1);
13         end
14         denominator=0;
15         for k=p+1:N,
16             denominator=denominator+P3(i,k-1);
17         end
18         A_new(i,j)=numerator/denominator;
19     end
20 end
21
22 % New priors estimation
23 c=P3(:,p+1);
```

Listing 5.9 MATLAB® ARMM iterative parameter estimation procedure.

```
1  function [A,c,alfa,sigma2] = ARMM_training(Lambda,m,p,MaxIter)
2
3  % Parameters initialization
4  c = rand(m,1);     % Priors vector
```

```
5  c = c/sum(c);       %
6  A = rand(m,m);       % Prob. state transition matrix
7  A = A./(sum(A,2)*ones(1,m));
8  sigma2 = var(Lambda); % Variance
9  alfa = zeros(p+1,m);  % Initialize autoreg. param.
10
11 for i=1:m,            % Stability is guaranteed
12     poles=rand(1,p);
13     eqx=poly(poles);
14     alfa(:,i)=[rand -eqx(2:end)].';
15 end
16
17 % Iteration...
18 for iter=1:MaxIter,
19     [P1,P2,~]=...
20     hamilton_algorithm(A,c,Lambda,alfa,sigma2);
21     P3=kim_algorithm(A,P1,P2,p);
22     [alfa,sigma2]=markov_ar_coef(p,Lambda,P3);
23     [A,c]=markov_ar_matx(p,A,P3);
24 end
```

The main application of ARMM is to describe the dynamics of time-series. Usually, one of the main objectives for the time-series model is to perform inference about the future. For this reason, the next section will be devoted to explain how to perform time-series predictions using the ARMM.

5.6 TIME SERIES PREDICTION WITH ARMM

After the training procedure presented in the previous section, it is assumed that the set of parameters for a particular ARMM is known. As we have seen, those parameters are essentially obtained by following an iterative optimization method that looks for model likelihood maximization over a given training data set. After having a fully functional ARMM, one of the main model purposes is to produce forecasts. That is, by using the observations history up to the present time to predict the value of future observations in an arbitrary time horizon.

With this in mind, the current section will be divided into two parts. In the first part, the one step ahead prediction will be addressed. Then, in the second part, the multiple steps ahead prediction will be the main subject. Regarding the latter, the h steps ahead prediction can be viewed as a closed-loop simulation for h time instants. In this simulation procedure, the observation values, predicted in previous time instants, are feedback into the model in order to generate further in time forecasts.

5.6.1 ONE STEP AHEAD TIME SERIES PREDICTION

Consider, once again, a p-order autoregressive model whose parameters are function of the actual active Markov state. Assuming that this state is $q_k = s_i$ the model is expressed as:

$$\lambda_{k|i} = \alpha_0^{s_i} + \alpha_1^{s_i}\lambda_{k-1} + \cdots + \alpha_p^{s_i}\lambda_{k-p} + e_k \qquad (5.79)$$

Furthermore, taking into consideration that we have prior knowledge about the observations values up to present time k, the objective is to use the above model in order to estimate the observation for the next time instant. That is, finding λ_{k+1} from the data set Λ_k.

If a m states ARMM is assumed, the answer to this problem requires the calculation of the expected value for λ_{k+1} conditioned to each and every model states. Then, the m different estimations are linearly combined to generate the one step ahead prediction. The predictions linear combination is done by a weight sum of the partial prediction obtained from each of the m different models. The weight factor associated to the ith prediction is equal to the probability of having, at time $k+1$, the ith state active. In particular, consider the $k+1$ observation value generated by the model associated to the state s_i. That is:

$$\lambda_{k+1} = \alpha_0^{s_i} + \alpha_1^{s_i}\lambda_k + \cdots + \alpha_p^{s_i}\lambda_{k-p+1} + e_{k+1} \qquad (5.80)$$

The expected value of λ_{k+1}, provided that the active state is s_i and assuming the knowledge on the observations history up to time instant k, is:

$$\mathfrak{E}\{\lambda_{k+1}|q_{k+1} = s_i \wedge \Lambda_k\} = \alpha_0^{s_i} + \alpha_1^{s_i}\lambda_k + \cdots + \alpha_p^{s_i}\lambda_{k-p+1} \qquad (5.81)$$

Observe that,

$$\mathfrak{E}\{\lambda_{k-j}|q_{k+1} = s_i \wedge \Lambda_k\} = \lambda_{k-j} \qquad (5.82)$$

for $j = 0, \cdots, p-1$.

Denoting $\mathfrak{E}\{\lambda_{k+1}|q_{k+1} = s_i \wedge \Lambda_k\}$ by $\hat{\lambda}_{k+1|i}$, the expression at (5.81) takes the form:

$$\hat{\lambda}_{k+1|i} = \alpha_0^{s_i} + \alpha_1^{s_i}\lambda_k + \cdots + \alpha_p^{s_i}\lambda_{k-p+1} \qquad (5.83)$$

The marginal expected value of λ_{k+1} must be computed considering all the possible active states at time instant $k+1$. In [12], it is shown that this value can be obtained by a weighting sum of $\hat{\lambda}_{k+1|i}$, for $i = 1, \cdots, m$,

235

where the weighting factors are the probabilities of each state to be active at time $k + 1$. That is,

$$\hat{\lambda}_{k+1} = \sum_{i=1}^{m} \hat{\lambda}_{k+1|i} \cdot P(q_{k+1} = s_i | \Lambda_k) \tag{5.84}$$

This equation suggests the necessity of having some procedures capable of predicting the active state at $k + 1$ from the knowledge of Λ_k. Since the active state at time k is known, the probability regarding the next active state can be easily obtained by looking at the probability transitions matrix $\mathbf{A}$. If the active state at time k is s_i, then the probability of getting s_j as the active hidden state at time instant $k + 1$ is just the value present at the ith row, j^{th} column of matrix $\mathbf{A}$. Let this element be denoted by a_{ij}. Hence,

$$a_{ji} = P(q_{k+1} = s_i | q_k = s_j) \tag{5.85}$$

Multiplying both terms of (5.85) by $P(q_k = s_j | \Lambda_k)$ yields:

$$P(q_k = s_j | \Lambda_k) \cdot a_{ji} = P(q_{k+1} = s_i | q_k = s_j) \cdot P(q_k = s_j | \Lambda_k) \tag{5.86}$$

By adding all the elements along the j coordinate then,

$$\sum_{j=1}^{m} P(q_k = s_j | \Lambda_k) \cdot a_{ji} = \sum_{j=1}^{m} P(q_{k+1} = s_i | q_k = s_j) \cdot P(q_k = s_j | \Lambda_k) \tag{5.87}$$

where the right hand term is equal to $P(q_{k+1} | \Lambda_k)$. For this reason, the previous expression can be rewritten as:

$$P(q_{k+1} = s_i | \Lambda_k) = \sum_{j=1}^{m} P(q_k = s_j | \Lambda_k) \cdot a_{ji} \tag{5.88}$$

Once again, the computation of $P(q_k = s_j | \Lambda_k)$ is carried out by the Hamilton algorithm presented in the previous section. Additionally, expression (5.84) can also be efficiently computed by describing it in vector form. First let's define the following vectors:

$$\hat{\boldsymbol{\lambda}}_{k+1} = [\hat{\lambda}_{k+1|1} \quad \cdots \quad \hat{\lambda}_{k+1|m}]^T \tag{5.89}$$

$$\Upsilon_{k+1,k} = [P(q_{k+1} = s_1 | \Lambda_k) \quad \cdots \quad P(q_{k+1} = s_m | \Lambda_k)] \tag{5.90}$$

where, generically, $\Upsilon_{i,j}$ represents the probabilities vector:

$$\Upsilon_{i,j} = \begin{bmatrix} P(q_i = s_1|\Lambda_j) & P(q_i = s_2|\Lambda_j) & \cdots & P(q_i = s_m|\Lambda_j) \end{bmatrix} \quad (5.91)$$

Taking these new vectors into consideration, the probability $P(q_{k+1} = s_i|\Lambda_k)$, for $i = 1, \cdots, m$, previously described at (5.88), can be represented by the matricial operation:

$$P(q_{k+1} = s_i|\Lambda_k) = \Upsilon_{k,k} \cdot \mathbf{A}_i \quad (5.92)$$

where $\mathbf{A}_i$ denotes the ith column of the transition probability matrix $\mathbf{A}$. Hence, the equality (5.84) can be described by:

$$\begin{aligned} \hat{\lambda}_{k+1} &= \Upsilon_{k+1,k} \cdot \hat{\lambda}_{k+1} \\ &= \Upsilon_{k,k} \cdot \mathbf{A}_i \cdot \hat{\lambda}_{k+1} \end{aligned} \quad (5.93)$$

and can be used to generate the one step ahead prediction based on the model and on a set of historical observations. Frequently, it is important to generate the predictions even further in the future. The following section addresses this problem and provides a framework to obtain long term forecasts.

5.6.2 MULTIPLE STEPS AHEAD TIME SERIES PREDICTION

This section provides the ARMM equations used to generate long term predictions. That is, knowing the ARMM model parameters and the observations historical background up to time k, the objective is to generate, for some future horizon $h \in \mathbb{N}$, the value of $\hat{\lambda}_{k+h|k}$.

In order to do this, let's first assume a particular situation and, from the observed pattern, derive the general equations that will allow the prediction for any arbitrary time horizon. Without loss of generality, let's start by assuming a third order ARMM with three hidden states. The observations are generated, assuming that the present active hidden state is $q_k = s_i$, by the following autoregressive model:

$$\lambda_{k|i} = \alpha_0^{s_i} + \alpha_1^{s_i}\lambda_{k-1} + \alpha_2^{s_i}\lambda_{k-2} + \alpha_3^{s_i}\lambda_{k-3} \quad (5.94)$$

for $i = \{1, 2\}$.

Suppose now that it is intended to generate the prediction, not one, but two steps ahead. One way to do it is by feeding back, the one step ahead prediction to the model. That is,

$$\hat{\lambda}_{k+2|i} = \alpha_0^{s_i} + \alpha_1^{s_i}\hat{\lambda}_{k+1} + \alpha_2^{s_i}\lambda_k + \alpha_3^{s_i}\lambda_{k-1} \quad (5.95)$$

where $\hat{\lambda}_{k+1}$ is computed from (5.84).

The marginal prediction value is obtained, using a linear combination similar to (5.84), by:

$$\hat{\lambda}_{k+2} = \sum_{i=1}^{m} \hat{\lambda}_{k+2|i} \cdot P(q_{k+2} = s_i | \Lambda_k) \tag{5.96}$$

Again, it is necessary to compute the probability $P(q_{k+2} = s_i | \Lambda_k)$. Since the Markov chain is homogeneous then,

$$a_{ji} = P(q_{k+2} = s_i | q_{k+1} = s_j) \tag{5.97}$$

By performing the same sequence of operations as for equation (5.85), yields,

$$P(q_{k+2} = s_i | \Lambda_k) = \sum_{j=1}^{m} P(q_{k+1} = s_j | \Lambda_k) a_{ji} \tag{5.98}$$

Replacing $P(q_{k+1} = s_j | \Lambda_k)$ by (5.84) leads to:

$$
\begin{aligned}
P(q_{k+2} = s_i | \Lambda_k) &= \sum_{j=1}^{m} a_{ji} \sum_{l=1}^{m} P(q_k = s_l | \Lambda_k) \cdot a_{lj} \\
&= \sum_{l=1}^{m} P(q_k = s_l | \Lambda_k) \sum_{j=1}^{m} a_{lj} a_{ji}
\end{aligned} \tag{5.99}
$$

which, for $i = 1$ and $m = 3$, gives:

$$
P(q_{k+2} = s_1 | \Lambda_k) = \sum_{l=1}^{3} P(q_k = s_l | \Lambda_k) \sum_{j=1}^{3} a_{lj} a_{j1}
$$

$$
\begin{aligned}
&= \sum_{l=1}^{3} P(q_k = s_l | \Lambda_k) \left[a_{l1} a_{11} + a_{l2} a_{21} + a_{l3} a_{31} \right] \\
&= P(q_k = s_1 | \Lambda_k) \left[a_{11} a_{11} + a_{12} a_{21} + a_{13} a_{31} \right] + \\
&\quad P(q_k = s_2 | \Lambda_k) \left[a_{21} a_{11} + a_{22} a_{21} + a_{23} a_{31} \right] + \\
&\quad P(q_k = s_3 | \Lambda_k) \left[a_{31} a_{11} + a_{32} a_{21} + a_{33} a_{31} \right]
\end{aligned} \tag{5.100}
$$

The above expression can be represented by the following matricial formulation:

$$P(q_{k+2} = s_1|\Lambda_k) =$$
$$\begin{bmatrix} P(q_k = s_1|\Lambda_k) & P(q_k = s_2|\Lambda_k) & P(q_k = s_3|\Lambda_k) \end{bmatrix} \cdot$$
$$\begin{bmatrix} a_{11}a_{11} + a_{12}a_{21} + a_{13}a_{31} \\ a_{21}a_{11} + a_{22}a_{21} + a_{23}a_{31} \\ a_{31}a_{11} + a_{32}a_{21} + a_{33}a_{31} \end{bmatrix} \tag{5.101}$$
$$= \Upsilon_{k,k} \cdot \mathbf{A} \cdot \mathbf{A}_1$$

In abstract, for an arbitrary hidden state s_i, one gets:

$$P(q_{k+2} = s_i|\Lambda_k) = \Upsilon_{k,k} \cdot \mathbf{A} \cdot \mathbf{A}_i \tag{5.102}$$

Notice that, since $\Upsilon_{k+2,k}$ is equal to:

$$\Upsilon_{k+2,k} = \begin{bmatrix} P(q_{k+2} = s_1|\Lambda_N) & \cdots & P(q_{k+2} = s_m|\Lambda_N) \end{bmatrix} \tag{5.103}$$

then, for the set of all possible hidden states, expression (5.102) can be described by [12]:

$$\Upsilon_{k+2,k} = \Upsilon_{k,k} \cdot \mathbf{A} \cdot \mathbf{A}$$
$$= \Upsilon_{k,k} \cdot \mathbf{A}^2 \tag{5.104}$$

It is straightforward to show that, for an arbitrary prediction horizon h, expression (5.104) can be extended to:

$$\Upsilon_{k+h,k} = \Upsilon_{k,k} \cdot \mathbf{A}^h \tag{5.105}$$

For this reason, and for a h steps ahead prediction horizon, the expression (5.96) results in the following matricial description:

$$\hat{\lambda}_{k+h} = \Upsilon_{k+h,k} \cdot \hat{\boldsymbol{\lambda}}_{k+h}$$
$$= \Upsilon_{k,k} \cdot \mathbf{A}^h \cdot \hat{\boldsymbol{\lambda}}_{k+h} \tag{5.106}$$

where the vector $\hat{\boldsymbol{\lambda}}_{k+h}$ is defined as:

$$\hat{\boldsymbol{\lambda}}_{k+h} = \begin{bmatrix} \hat{\lambda}_{k+h|1} & \hat{\lambda}_{k+h|2} & \cdots & \hat{\lambda}_{k+h|m} \end{bmatrix}^T \tag{5.107}$$

To conclude, the h steps ahead prediction, for $h > 1$ is different from the one step ahead prediction in the sense that now there is a feedback path between the model input and the past generated predictions. It is important to notice that, by feeding back predictions and not the true values, the prediction error will increase with the increased prediction horizon value.

Finally, taking into consideration the set of equations derived so far, the long term prediction can be obtained, recurrently, by executing the following steps:

For each k value execute the following steps sequence:

1st **step:** Compute $\Upsilon_{k,k}$ from the Hamilton algorithm. The regressors vector depends on the present value of i and on the autoregressive model order p. This is true since, if $i \leq p$, the computation of $\lambda_{k+i|j}$ must be done using:

$$\hat{\lambda}_{k+i|j} = \alpha_0^{s_j} + \sum_{l=1}^{t} \alpha_l^{s_j} \hat{\lambda}_{k+i-l} + \sum_{l=t+1}^{p} \alpha_l^{s_j} \lambda_{k+i-l} \tag{5.108}$$

where $t = p - i + 1$. On the other hand, if $i > p$ one should use in alternative:

$$\hat{\lambda}_{k+i|j} = \alpha_0^{s_j} + \sum_{l=1}^{p} \alpha_l^{s_j} \hat{\lambda}_{k+i-l} \tag{5.109}$$

2nd **step:** From $i = 1$ to h execute the following computations:
(a) Generate the regressors vector matrix $\mathbf{H}_{k+i}$;
(b) Compute the conditional predictions vector $\hat{\lambda}_{k+i}$ using the vectorial organization of $\hat{\lambda}_{k+i|j}$ for $j = 1, \cdots, m$ where $\hat{\lambda}_{k+i|j} = \mathbf{H}_{k+i} \cdot \hat{\mathbf{c}}_{s_i}$;
(c) Obtain the value for the marginal prediction from:

$$\hat{\lambda}_{k+i} = \Upsilon_{k,k} \cdot \mathbf{A}^i \cdot \hat{\lambda}_{k+i}$$

The MATLAB® code, presented in Listing 5.10, implements the above time-series prediction procedure. This function calls two other sub functions presented in Listings 5.11 and 5.12. The former is responsible to build the regressors matrix and the later the conditional probabilities vector. Observe also, at the beginning of Listing 5.10, a call to the Hamilton algorithm described in Section 5.5.

Listing 5.10 MATLAB® ARMM h step ahead prediction.

```
1  function [L_pred,L_meas] = ARMM_kstepahead(Lambda,A,c,alfa,sigma2,h)
2
3  % Parameters initialization
4  [p,m] = size(alfa);          % Number of states and
5  p = p-1;                     % AR order
6  N=length(Lambda);            % Number of observations
7  L_pred=zeros(N,h);           % Predicted obs. matrix
8  L_meas=zeros(N,h);           % Measured obs. matrix
9
10 % Execution (1st step)
11 ["̃,P2,̃]=...
12 hamilton_algorithm(A,c,L_pred,alfa,sigma2);
13 for k=p+1:N-h,
14     Upsilon_k_k=P2(:,k).';
15     % (2nd step)
16     for i=1:h,
17         H_k_i=...
18         regressors_matrix(L_pred,Lambda,k,p,i);
19         vec_lambda=conditional_lambda(H_k_i,alfa,m);
20         L_pred(k,i)=Upsilon_k_k*(A^h)*vec_lambda;
21         L_meas(k,i)=Lambda(k+i-1);
22     end
23 end
```

Listing 5.11 Function called by the MATLAB® Listing 5.10. Returns the regressors vectors matrix.

```
1  function H=regressors_matrix(Lambda_pred,Lambda,k,p,i)
2
3  if i<=p
4      H=[1 Lambda_pred(k,i-1:-1:1) Lambda(k:-1:k-(p-i))];
5  else
6      H=[1 Lambda_pred(k,i-1:-1:i-p)];
7  end
```

Listing 5.12 Function called by the MATLAB® Listing 5.10. Returns the conditional probabilities vector.

```
1  function lambda=conditional_lambda(H,alfa,m)
2
3  lambda=zeros(m,1);
4  for k=1:m,
5      lambda(k)=H*alfa(:,k);
6  end
```

5.7 SUMMARY

A review of the autoregressive Markov models, in the context of time-series modelling and prediction, was the main objective of this chapter. As noted, this type of models relies on two distinct components. One that captures the data time structure and the other that implements an automatic switching paradigm based on Markov chains. The first element relies on a set of autoregressive models in order to tackle the situation where observations are not conditionally independent of one another. This fact leads to the ability to use ARMM in problems where the ordinary hidden Markov models cannot be used. The time-series prediction problem is one of those cases.

In addition to the theoretical aspects of this theme, illustrative computer code was provided within the text. By pairing both the mathematical description and its computer representation, one believes that the subject can be more easily grasped.

It's worth to note that, inline with the rest of the book, code vectorization was relaxed and replaced by a more readable "if-then-else" format. That is, the code's pedagogical nature superseded the quest for computational efficiency. In order to illustrate this statement, please compare the MATLAB® code presented in Listing 5.4 with the one in Listing 5.13. Both refer to the Hamilton algorithm. However, the latter is a vectorized version of the former. The same can be said about the code presented in Listing 5.5 and the one in Listing 5.14. In this case, the MATLAB® functions regard the Kim's algorithm.

As can be seen, the vectorized version, even if more computationally efficient, is less intuitive since it lacks the structured computer language that we humans are able to handle more easily.

Listing 5.13 A more vectorized MATLAB® version for Hamilton's algorithm.

```
1  function [P1,P2,lik]=hamilton_algorithm_vect(A,c,Lambda,alpha,sigma2)
2
3  Lambda=Lambda(:); % Just to ensure that Lambda is a column vector...
4  [p,m]=size(alpha);
5  p=p-1;
6
7  % Observations number.
8  N=length(Lambda);
9
10 % Space allocations
11 eta=zeros(m,N);P1=zeros(m,N);P2=zeros(m,N);
12
13 % For each observation compute eta_k for k=p+1,...,N
14 for k=p+1:N
```

```
15    for i=1:m,
16        h_k=[1;Lambda(k-1:-1:k-p)];
17        eta(i,k)=(1/(sqrt(2*pi*sigma2)))*...
18            exp(-((Lambda(k)-h_k'*alpha(:,i)).^2)/(2*sigma2));
19    end
20 end
21
22 % Initialize xi
23 k=p+1;
24 for i=1:m,
25     P1(i,k)=c(i);
26 end
27
28 % Compute P2
29 for k=p+1:N-1
30     P2(:,k)=(P1(:,k).*eta(:,k))/(ones(1,m)*(P1(:,k).*eta(:,k)));
31     P1(:,k+1)=A'*P2(:,k);
32 end
33 P2(:,N)=(P1(:,N).*eta(:,N))/(ones(1,m)*(P1(:,N).*eta(:,N)));
34
35 % Log-Likelihood...
36 lik=0;
37 for k=p+1:N,
38     lik=lik+log(ones(1,m)*(P1(:,k).*eta(:,k)));
39 end
```

Listing 5.14 A more vectorized MATLAB® version for Kim's algorithm.

```
1 function P3=kim_algorithm_vect(A,P1,P2,p)
2
3 [m,N]=size(P1);
4 P3=zeros(m,N);
5
6 % Initialization...
7 P3(:,N)=P1(:,N);
8
9 % Recursion...
10 for k=N-1:-1:p+1,
11     P3(:,k)=P1(:,k).*(A*(P3(:,k+1)./P2(:,k+1)));
12 end
```

The next chapter presents some illustrative examples on using hidden Markov models. In particular, all the distinct models presented within this book will be illustrated by means of case studies. It is worth to highlight that all the results were obtained using the MATLAB® functions developed along the book.

6 Selected Applications

"As a rule, software systems do not work well until they have been used, and have failed repeatedly, in real applications."

–Dave Lorge Parnas

Hidden Markov models have found application in a myriad of different areas. For example in molecular biology, speech synthesis and recognition, stock prices prediction, gesture recognition and in many biomedical applications such as epileptic seizure modelling, electrocardiogram analysis and many more. It is not the objective of the current book to address any of those applications in detail. As referred in the introductory chapter, this book aims to present the hidden Markov model paradigm and at the same time, to translate it into computer code. That is, the materials presented in the previous chapters are broad in scope and do not focus on any specific application. However, this text can not be complete without exploring a couple examples where some of the algorithms and techniques introduced during the past five chapters can be applied.

In order to carry out this task, two distinct cases will be addressed. The first example describes a biomedical application. In this case a continuous distribution hidden Markov model will be used to classify the results of cardiotocography signals into three different classes: "normal", "suspect" and "pathological". The second example describes a time-series problem where the solar radiation is modelled using an autoregressive Markov model. Using this model, both short- and long-term forecasts are performed.

6.1 CARDIOTOCOGRAPHY CLASSIFICATION

Cardiotocography is a medical exam, used during the last trimester of pregnancy, to monitor simultaneously the mother uterus contractions and the fetal heart rate. This exam is performed by attaching two different transducers to the mother's abdomen[1]: an ultrasonic sensor and a load cell. Additionally, a push-button is used as an event detector. The ultrasound unit has a disk-like appearance and is placed over the uterus area. To improve the waves propagation, a conductive gel is spread between the skin and the transducer. This ultrasound sensor provides information regarding the movement of the fetus's heart. At each heart beat, the changes in frequency of the ultrasound waves are detected and the cardiotocograph can then calculate the fetal heart rate. To monitor uterine contractions, an external dynamometer is attached around the pregnant abdomen by means of a belt. The uterus contraction magnitude is proportional to the strain exerted on the load cell. Finally, the mentioned push-button must be pressed manually by the pregnant woman whenever she feels fetal movement.

The observation is carried out during a time interval that can span from twenty minutes to one hour. At the end of the exam, the data provided by the cardiotocograph machine is represented as a pair of line plots over a scale paper sheet. Figure 6.1 shows an example of such signals.

The upper plot provides information about the fetus's heart rate, in beats per minute, and the lower plot the uterine contractions in a normalized scale.

The exam diagnostics is made by visual inspection regarding the two plots profile. Following some standard guidelines, and by using its own personal experience, the health technician can infer about the exam outcome. That is, if it is normal or if some unusual characteristics were detected.

Since the exam diagnostic strongly depends on the examiner skills, in the last few years several decision support strategies based on machine learning techniques have been proposed. For example [10], [35], [36], [30], [33] and [39], just to name a few. The common objective to all these works is to harmonize the cardiotocography interpretation and make it less prone to human factors.

In this example, an alternative classification strategy is addressed. Three continuous observation hidden Markov models will be used to classify the data provided by a set of cardiotocography medical exams into

---

[1]Alternatively the transducers can be placed inside the uterus and into the fetus scalp.

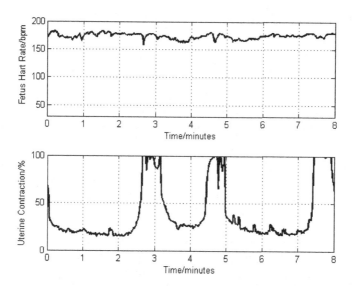

Figure 6.1 Excerpt of cardiotocography signals found during a medical exam. The top signal relates to the fetus heart rate frequency and the bottom to the uterine contractions.

three different classes: "normal", "suspect" and "pathological". Each class will be associated with one of the hidden Markov models where its parameters Θ are estimated using the Baum-Welch method over given data sets. Without loss of generality, in this section all the models are assumed to have the same structure: four hidden states and probability density functions composed by a mixture of four Gaussian functions.

The data used in this section was downloaded from the *Machine Learning Repository* web site and is available on-line, free of charge, from [25]. The data file is composed by more than two thousand entries. Each entry represents a different patient data whose diagnostic belongs to one of the three above mentioned categories. Furthermore, the available data file is composed, not just by the two raw cardiotocograph signals, but also by a set of twenty one features computed from the measured signals. The first feature is the fetal heart beat baseline, the second the number of heart beat accelerations, the third the number of uterine contraction per second and so on. An exhaustive list regarding all the existing features, along with their description and data processing techniques, can be found in [6].

In the present approach only the first seven features will be used. Hence, each observation presented to the model, λ, is a vector with dimension seven. Moreover, the following procedure sequence will be carried out:

- First, using the Baum-Welch method iterated over 50 times, three distinct continuous observations hidden Markov models will be trained. One for each possible diagnosis class and using a set of 150 examples;
- Second, a total of 75 validation examples, composed by 25 examples of each of the three possible classes, are applied sequentially to each hidden Markov model. The value of the log-likelihood regarding each model is then computed.

The data classification, as belonging to one of the three possible outcomes, will depend on which of the three hidden Markov model provides the higher likelihood value for a given example.

The MATLAB® code presented in Listing 6.1 implements the above execution strategy. First, the estimation and validation data is loaded from a file, which contains six matrices. Three, with dimension 7×150, containing estimation data and denoted by "normal", "suspect" and "pathological". The other three, with dimensions 7×25 designated by "VNormal", "VSuspect", and "VPathological", that handle the validation data.

The computer code in Listing 6.2 describes the main MATLAB® function. Its purpose is to return a continuous hidden Markov model trained with the Baum-Welch method. On the other hand, the MATLAB® function presented in Listing 6.3 returns a symmetric positive definite matrix whose entries are generated by a normally distributed process.

Listing 6.1 Cardiotocography classification MATLAB® script.

```
1  % Clear workspace and load data
2  clear all;
3  close all;
4  clc;
5
6  load reducedctgdata.mat
7
8  % Define model structure
9  m = 4;              % Number of hidden states
10 M = 4;              % Number of Gaussian per mixture
11 MaxIter = 50;       % Number of iterations
12
13 % Model Estimation
14 normalHMM     = ctg_model(Normal,m,M,MaxIter);
15 suspectHMM    = ctg_model(Suspect,m,M,MaxIter);
```

```
16 pathologicHMM = ctg_model(Pathological,m,M,MaxIter);
17
18 % Model Validation
19 N=length(VNormal);
20 ResNormal=zeros(N,2);
21 for i=1:N,
22     LL = [loglik(VNormal(:,i),normalHMM), ...
23           loglik(VNormal(:,i),suspectHMM), ...
24           loglik(VNormal(:,i),pathologicHMM)];
25     [~,inx] = max(LL);
26     ResNormal(i,1)=1;ResNormal(i,2)=inx;
27 end
28
29 N=length(VSuspect);
30 ResSuspect=zeros(N,2);
31 for i=1:N,
32     LL = [loglik(VSuspect(:,i),normalHMM), ...
33           loglik(VSuspect(:,i),suspectHMM), ...
34           loglik(VSuspect(:,i),pathologicHMM)];
35     [~,inx] = max(LL);
36     ResSuspect(i,1)=2;ResSuspect(i,2)=inx;
37 end
38
39 N=length(VPathologic);
40 ResPathological=zeros(N,2);
41 for i=1:N,
42     LL = [loglik(VPathological(:,i),normalHMM),...
43           loglik(VPathological(:,i),suspectHMM), ...
44           loglik(VPathological(:,i),pathologicHMM)];
45     [~,inx] = max(LL);
46     ResPathological(i,1)=3;ResPathological(i,2)=inx;
47 end
```

Listing 6.2 Initialize model parameters and apply the Baum-Welch method.

```
1 function HMM = ctg_model(Data,m,M,MaxIter)
2
3 % Markov Model
4 [d,~]=size(Data);      % Observations dimension
5
6 % Hidden Markov Model for class = Normal
7 % Parameters initialization
8 c=rand(m,1);           % Priors
9 c=c/sum(c);            %
10 A=rand(m,m);          % Hidden state probabilities
11 A=A./(sum(A,2)*ones(1,m));
12 W=rand(m,M);          % Mixture weights
13 W=W./(ones(m,1)*sum(W));
14 mMu=(ones(M,1)*mean(Data,2).');
15 Mu=zeros(m,M,d);
16
17 for i=1:m,
18     Mu(i,:,:)=mMu+randn(M,d);
19 end
20
21 mSigma=cov(Data.');
22 Sigma=zeros(m,M,d,d);
23
24 for i=1:m,
```

```
25    for j=1:M,
26        Sigma(i,j,:,:)=mSigma+randn_cov(d,0);
27    end
28 end
29
30 % Train hidden Markov model
31 [A,Mu,Sigma,W,c]=baum_welch(Data,A,Mu,Sigma,W,c,MaxIter);
32
33 % Register model parameters
34 HMM.A = A;
35 HMM.Mu = Mu;
36 HMM.Sigma = Sigma;
37 HMM.W = W;
38 HMM.c = c;
```

Listing 6.3 Generate a normally distributed random symmetric and positive definite matrix.

```
1 function A = randn_cov(n,eigen_mean)
2
3 q = randn(n,n);
4 A = q' * diag(abs(eigen_mean+randn(n,1))) * q;
```

After a set of tryouts, the best set of generated Markov models leads to a success rate of around 70%. For the "Normal" validation set the system correctly classifies 19 of the total 25 cases. The remain 6 cases are classified as "Suspect". For the "Suspect" validation data set the system is correct in 15 of the 25 examples. The wrong classifications are distributed as follows: 2 for "Normal" and 8 for "Pathological" classes. Finally, the use of the "Pathological" validation data set led to a classification success of 18 over the 25 examples. The remaining misclassifications are all attributed to the "Normal" class.

6.2 SOLAR RADIATION PREDICTION

Currently, a considerable part of the world's agricultural production takes place inside greenhouses. For example China, the major indoor grower in the world, has a total installed area over 2.700.000 ha [18]. This indoor cultivation strategy has the advantage to enable crop growth adjustment by artificially modifying both the environmental conditions and the plants nutrition.

The main goal of indoor agriculture is to optimise the balance between the production economic return and the operation costs of the climate actuators such as ventilation, heating and so on. The energy consumption to maintain the appropriate indoor climate, and eventually actuators wear and tear, can be reduced by designing suitable indoor control system strategies.

State-of-the-art greenhouse climate controllers are based on models to simulate and predict greenhouse environment behaviour and then, this information is used to compute the required actuation signals. These models must be able to accurately describe the indoor climate process dynamics, which is a function of the control actions taken, the outside climate and the thermal and optic properties of the building. Moreover, if predictive or feed-forward control techniques are to be applied in the computation of the required actuator signals, it is necessary to employ models to describe and predict the indoor and outdoor environmental conditions such as the air temperature and the solar radiation. From all the physical variables that play an important role in the greenhouse indoor climate, solar radiation is the one that has the biggest impact in the heat load.

For this reason, the current case study regards the modelling and prediction of the solar radiation. In this case, it is assumed that the dynamic system state space is not directly observable. Hence, the solar radiation dynamic behaviour will be described considering a time-series model structure. In this context, the appropriate response, in a particular time instant, will depend exclusively on past observations.

The solar radiation, viewed from the time-series perspective, is a very complex modelling and prediction problem due to the extremely uncorrelated high frequency components. Although the low-frequency profile could be obtained from a deterministic solar radiation model, the high frequency oscillation, due to disturbances such as clouds and atmosphere attenuation, is extremely difficult to predict using only past information. In general, for a given time series model, the evolution of the relative prediction error versus the forecast horizon, will show a logarithmic profile with the error rate rising fast in the first thirty or forty steps ahead of the prediction.

The objective of this section is not to undergo the full problems associated with the solar irradiation prediction. Instead, the main purpose is to have an example where it is possible to test the code derived in the previous Chapter. With this objective in mind, the intention is to perform accurate predictions of the solar radiation along a particular day of the year. The predictions will be conducted aiming for two different prediction horizons: one step ahead and ten steps ahead.

Before proceeding to the simulations, let's refer back to the fact that the observations used in this section were acquired, using a pyranometer located at the North of Portugal. A pyranometer is a device used to measure the solar irradiance on a planar surface. The data provided is the solar radiation flux density, measured in (W/m^2), within a wavelength interval $[0.3, 3]$ μm. The data was acquired using a sampling period of one minute which leads to a total of 1440 samples per day.

The tests performed, and presented below, have used the data from four consecutive days chosen due to their high irradiation fluctuations. Figure 6.2 shows the shape of the used data.

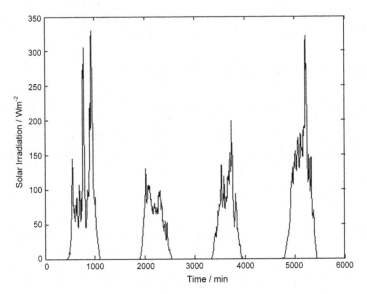

Figure 6.2 Solar irradiation data used for time series prediction with ARMM.

The data concerning the first three days has been used for parameter estimation while the solar radiation predictions were done on the fourth day.

By visual inspection of the data presented in Figure 6.2, it is possible to detect at least two different system regimes. A day regime, where the solar radiation is greater than zero, and a night regime where the solar radiation is zero. However, additional regimes can be further considered. For example, during the day there is a region where the signals first derivative is increasing in average. This tendency will be maintained from sunrise up to the Sun's peak hour. After that, the signal will show a decreasing tendency ending at sunset. In this context, it is possible to define a night regime, an ascent regime and a descent regime. For this reason, in this case study, a three hidden states autoregressive Markov model will be used to model the solar radiation dynamics. Moreover, it is assumed a fourth order autoregressive model associated with each regime. However, this doesn't need to be so and distinct autoregressive orders can be used instead. For example, regimes with simpler dynamics can be represented by lower order models. Nevertheless, for simplicity reasons, it is assumed equal autoregressive model orders for all regimes. With these three hidden states, it is expected that each one of the autoregressive models handles a different data segment.

The autoregressive Markov model training procedure, and the time-series prediction results, were computed by the MATLAB® script in Listing 6.4.

Listing 6.4 MATLAB® script for solar irradiation prediction.

```
1  clear all;
2  close all;
3  clc;
4
5  % Load matrix with 1440 x 361 of measured radiation
6  load('solardata.mat');
7
8  % Parameters Estimation data
9  Lambda = [sol.day(:,150);sol.day(:,151);sol.day(:,152)].';
10
11 % Constants
12 MaxIter=100; % maximum number of training iterations
13 m=3;         % number of hidden states
14 p=4;         % AR model order
15 h=10;        % prediction horizon
16
17 % Training
18 [A,c,alfa,sigma2]=ARMM_training(Lambda,m,p,MaxIter);
19
20 % Prediction
21 Lambda = solar.day(:,153).'; % Validation Data
22 [Lambda_pred,Lambda_meas] = ...
23     ARMM_kstepahead(Lambda,A,c,alfa,sigma2,h);
24
25 % Results
26 figure(1);
27 plot([Lambda_meas(:,1) Lambda_pred(:,1)])
28 axis([0 1440 0 350])
29 xlabel('Time/min');
30 ylabel('Solar Irradiation/Wm^{-2}');
31 legend('Measured','one step ahead prediction');
32 error=Lambda_meas(:,1)-Lambda_pred(:,1);
33
```

253

```
34  % Compute prediction error
35  seq1=sqrt((error'*error)/length(error));
36  figure(2);
37  plot([Lambda_meas(:,h) Lambda_pred(:,h)])
38  axis([0 1440 0 400])
39  xlabel('Time/min');
40  ylabel('Solar Irradiation/Wm^{-2}');
41  legend('Measured',[num2str(h) ' step ahead prediction']);
42  y=Lambda_meas(:,h);ye=Lambda_pred(:,h);
43  error=Lambda_meas(:,h)-Lambda_pred(:,h);
44
45  % Compute prediction error
46  seq60=sqrt((error'*error)/length(error));
```

After executing the above MATLAB® script, the one step ahead and ten step ahead predictions were obtained. Figure 6.3 presents the overall prediction aspect for the one step ahead prediction and Figure 6.4 regards the result obtained from ten steps ahead prediction.

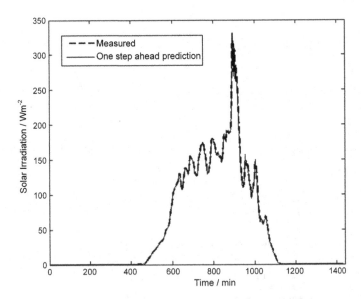

Figure 6.3 One step ahead prediction using an ARMM.

In general, the autoregressive Markov model performs well in describing the data dynamic behaviour. For the one step ahead prediction the average prediction error is around $2W/m^2$ and for the ten steps ahead the prediction error is slightly over $14W/m^2$.

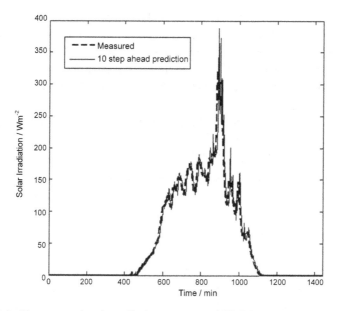

Figure 6.4 Ten steps ahead prediction using an ARMM.

6.3 SUMMARY

This chapter aims to present how the MATLAB® code, provided in the previous chapters, can be used to solve some particular problems in two distinct areas. The first example illustrates the behaviour of continuous hidden Markov models applied to a biomedical classification problem. The addressed problem regards the automatic discrimination of cardiotocographies into three different classes: "Normal", "Suspect" and "Pathological". Even without putting much effort in searching for the ideal model structure, it was possible to achieve an accuracy rate of around 70%.

The second example concerns the problem of time-series prediction. In this case study, the solar radiation is viewed as an example of a time-series and the aim is to predict it using an autoregressive Markov model. These types of models rely on two distinct components. One that captures the data time structure and the other that implements an automatic switching paradigm based on Markov chains. The first component is based on a set of autoregressive models able to handle the situation where observations are not conditionally independent one of each other.

The obtained solar radiation model has been used to perform predictions within two distinct forecast horizons. A short term prediction of one step ahead and a longer term horizon of ten steps ahead. From the obtained results it is possible to conclude that ARMM was able to provide very satisfactory results despite the complex nature of the proposed time-series.

As referred to in the beginning of this chapter, the universe of problems where hidden Markov models can be applied, is enormous. It is possible to find examples of their use in many and distinct disciplinary areas. It is now up to the readers to adapt the algorithms provided in this book to their particular research problem.

References

1. C. Aggarwal and C. Reddy. *Data Clustering: Algorithms and Applications.* Chapman & Hall, Boca Raton, USA, 2013.
2. L. Baum and T. Petrie. Statistical inference for probabilistic functions of finite state Markov chains. *Ann. Math. Statist.,* 37: 1554–1563, 1966.
3. L. Baum, T. Petrie, G. Soules and N. Weiss. A maximization technique occurring in the statistical analysis of probabilistic functions of Markov chains. *Ann. Math. Statist.,* 41: 164–171, 1970.
4. L. E. Baum and J. A. Egon. An inequality with applications to statistical estimation for probabilistic functions of a Markov process and to a model for ecology. *Bull. Amer. Meteorol. Soc.,* 73: 360–363, 1967.
5. M. Bazaraa, D. Sherali Hanif and C. M. Shetty. *Nonlinear Programming: Theory and Algorithms.* John Wiley & Sons, Inc., Hoboken, USA, 2006.
6. D. Ayres de Campos, J. Bernardes, A. Garrido, J. Marques de Sa and L. Pereira-Leite. Sisporto 2.0: A program for automated analysis of cardiotocograms. *Journal of Maternal-Fetal and Neonatal Medicine,* 9(5): 311–318, 2000.
7. R. Durbin, S. Eddy, A. Krogh and G. Mitchison. *Biological Sequence Analysis.* Cambridge University Press, Cambridge, UK, 1998.
8. R. A. Ficher. *Statistical Methods and Scientific Inference.* Hafner, London, UK, 1956.
9. R. Galbraith and T. Oenning. Iterative detection read channel technology in hard disk drives. Technical Report, HGST—A Wester Digital Company, 2008.
10. G. G. Georgoulas, C. D. Stylios, G. Nokas and P. Groumpos. Classification of fetal heart rate during labour using hidden Markov models. In *Neural Networks, 2004. Proceedings. 2004 IEEE International Joint Conference on,* 3: 2471–2475. IEEE, 2004.
11. J. D. Hamilton. Analysis of time series subject to changes in regime. *Journal of Econometrics,* 45: 39–70, 1990.
12. J. D. Hamilton. *Time Series Analysis.* Princeton University Press, Princeton, USA, 1994.
13. F. Jelinek. Continuous speech recognition by statistical methods. *Proceedings of the IEEE,* 64(4): 532–556, 1976.
14. F. Jelinek. *Statistical Methods for Speech Recognition.* MIT Press, Cambridge, USA, 1997.

15. B. Klemens. Probability vs. likelihood, July 2009.

16. J. Lember and A. Koloydenko. The adjusted viterbi training for hidden Markov models. *Bernoulli*, 14(1): 180–206, 2008.

17. I. MacDonald and W. Zucchini. *Hidden Markov and other Models for Discrete-valued Time Series*. Chapman and Hall, Boca Raton, USA, 1997.

18. R. Nair and S. Barche. Protected cultivation of vegetables—present status and future prospects in India. *Indian Journal of Applied Research*, 4(6): 245–247, 2014.

19. LAN/MAN Standards Committee of the IEEE Computer Society, editor. *Wire-less LAN Medium Access Control (MAC) and Physical Layer (PHY) specifications: High-speed Physical Layer in the 5 GHZ Band*, Chapter 11. The Institute of Electrical and Electronics Engineers, 1999.

20. J. Parks and I. Sandberg. Universal approximation using radial-basis-function networks. *Neuro Computation*, 3: 246–257, 1991.

21. A. Poritz. Linear predictive hidden Markov models and the speech signal. *ICASSP-82*, 1291–1294, 1982.

22. A. B. Poritz. Hidden Markov models: A guided tour. *IEEE Internationl Conference on Acoustic Speech and Signal Processing*, 0: 7–13, 1988.

23. L. Rabiner and B. Juang. An introduction to hidden Markov models. *IEEE ASSP Magazine*, 3(1): 4–16, January 1986.

24. S. S. Rao. *Engineering Optimization: Theory and Practice*. John Wiley & Sons, Inc., Hoboken, USA, 1996.

25. Machine Learning Repository. Cardiotocography data set, 2014.

26. T. Rolf. Direct maximization of the likelihood of a hidden Markov model. *Computational Statistics & Data Analysis*, 52(9): 4147–4160, May 2008.

27. J. B. Rosen. The gradient projection method for nonlinear programming. Part I. Linear constraints. *Journal of the Society for Industrial and Applied Mathematics*, 8(1): 181–217, 1960.

28. C. Shannon and W. Weaver. *The Mathematical Theory of Communication*. The University of Illinois Press, Illinois, USA, 1971.

29. J. A. Snyman. *Practical Mathematical Optimization: An Introduction to Basic Optimization Theory and Classical and New Gradient-based Algorithms*. Springer, New York, USA, 2005.

30. M. C. Sundar and G. Geetharamani. Classification of cardiotocogram data using neural network based machine learning technique. *International Journal of Computer Applications*, 47: 19–25, 2012.

31. W.H. Press, S.A. Teukolsky, W.T. Vetterling and B.P. Flannery. *Numerical Recipes: The Art of Scientific Computing*. Cambridge University Press, New York, USA, 2007.

32. S. Theodoridis and K. Koutroumbas. *Pattern Recognition*. Academic Press, London, UK, 2006.

33. P. Tomáš, J. Krohova, P. Dohnalek and P. Gajdoš. Classification of cardiotocography records by random forest. In *Telecommunications and Signal Processing (TSP), 2013 36th International Conference on*, 620–623. IEEE, 2013.

34. A. J. Viterbi. Error bounds for convolutional codes and an asymptotically optimum decoding algorithm. *IEEE Transactions on Information Theory*, 13(2): 260–269, 1967.

35. P. Warrick, E. F. Hamilton, R. E. Kearney and D. Precup. Classification of normal and hypoxic fetuses using system identification from intra-partum cardiotocography. In *Workshop on Machine Learning for Health-care Applications*, 2008.

36. P. Warrick, E. F. Hamilton, R. E. Kearney and D. Precup. Classification of normal and hypoxic fetuses from systems modeling of intrapartum cardiotocography. In *IEEE Trans. Biomed. Eng.*, 57: 771–779, 2010.

37. L. Welch. Hidden Markov models and the baum-welch algorithm. *IEEE Information Theory Society Newsletter*, 53:4, 10–13, 2003.

38. C. Wu. On the convergence properties of the EM algorithm. *Annals of Statistics*, 11: 95–103, 1983.

39. E. Yilmaz and C. Kilikcier. Determination of fetal state from cardiotocogram using ls-svm with particle swarm optimization and binary decision tree. *Computational and Mathematical Methods in Medicine*, 2013.

40. W. Zucchini and I. MacDonald. *Hidden Markov Models for Time Series—An Introduction using R*. CRC Press, Boca Raton, USA, 2009.

Index

Chapter 3

Fig. 3.21, p. 116

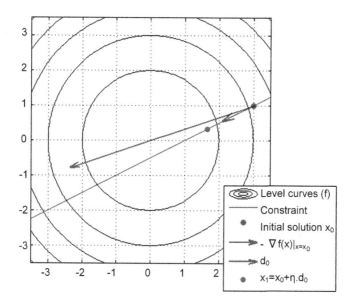

Chapter 3

Fig. 3.28, p. 130

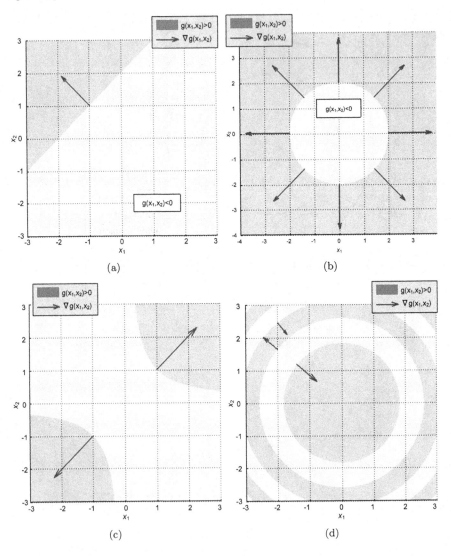

Chapter 3
Fig. 3.29, p. 131

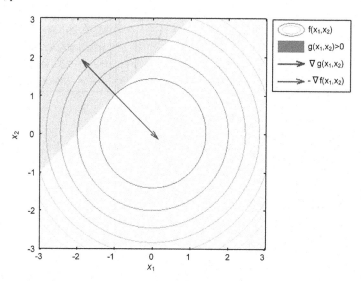

Chapter 3
Fig. 3.30, p. 131

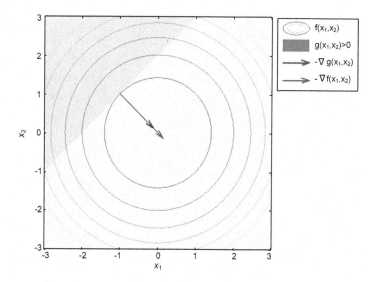

Chapter 3

Fig. 3.31, p. 133

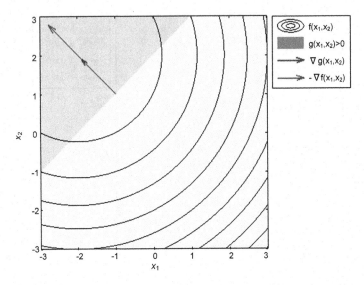

Printed in the United States
by Baker & Taylor Publisher Services